Molecular Cuisine

Twenty Techniques
Forty Recipes

Anne Cazor
& Christine Liénard

Molecular Cuisine

Twenty Techniques Forty Recipes

Photographs by Julien Attard
Translation by Gui Alinat
Foreword by Hervé This

CRC Press
Taylor & Francis Group
Boca Raton London New York

CRC Press is an imprint of the
Taylor & Francis Group, an **informa** business

CRC Press
Taylor & Francis Group
6000 Broken Sound Parkway NW, Suite 300
Boca Raton, FL 33487-2742

© 2012 by Taylor & Francis Group, LLC
CRC Press is an imprint of Taylor & Francis Group, an Informa business

No claim to original U.S. Government works

Printed in the United States of America on acid-free paper
Version Date: 20110829

International Standard Book Number: 978-1-4398-7163-8 (Hardback)

Visit the Taylor & Francis Web site at
http://www.taylorandfrancis.com

and the CRC Press Web site at
http://www.crcpress.com

Contents

Contents

Foreword
Hervé This

There is a lot of confusion among gourmands: that trendy "molecular gastronomy" is generally considered mistakenly as a new culinary trend, with chefs using liquid nitrogen, making alginate pearls, etc., or as a game for foodies, or even sometimes as a way to "put some chemistry in our food." The most innovative of us praise this renovation of cooking, whereas the most conservative fear that this new way of cooking will kill their favorite food, roasted turkeys, coq au vin, soufflés...

However, both sides are wrong, as molecular gastronomy is not cooking, but rather a scientific activity!

Indeed, "gastronomy" does not mean good food or elaborated cuisine, as many years of confusion have led some people to think. It is worth telling that the word "gastronomy" was introduced in French in 1801 by the poet Joseph Berchoux, and it was popularized two decades later (1825 precisely) by a gourmand lawyer, Jean-Anthelme Brillat-Savarin, in his *Physiology of Taste*. The book had a tremendous and lasting success all over the world, probably because it was cheering gourmandise, and also because it was a very talented literary piece.

Moreover Brillat-Savarin had the real talent of clearly defining "gastronomy" as "the reasoned knowledge concerning all aspects of man's food." For example, when one studies the history of the mayonnaise sauce, he or she is doing "historical gastronomy"; the study of the differences among cassoulets from various locations belongs to "geographical gastronomy;" the study of food habits is sociological gastronomy, etc.

It should now seem clear that the chemical and physical study of cooking could be called chemical gastronomy and physical gastronomy. But in 1988, with my old friend Nicholas Kurti, we proposed to call it "molecular gastronomy" because we had in mind the rapidly developing "molecular biology," in which the purpose and methods were similar. Indeed, we knew well that food science had been active for a long time, but it was also true that this food science did not consider culinary transformations. There was no study of soufflés, pheasants à l'Albufera, Black Forest cake, etc. We knew that science means "looking for the mechanisms of phenomena" using the scientific method, and we also knew that many interesting phenomena occur during the culinary activities, so we thought that it would be interesting and fair to create a new scientific discipline that would not be interested in the chemistry of food ingredients, or in industrial processes, but rather would focus on culinary transformations. We called it molecular gastronomy. As a first conclusion, let us repeat, here, that molecular gastronomy is a scientific discipline.

Well before 1988, Nicholas Kurti and I were differently interested in promoting a modernization of culinary practices. Being a physicist, Nicholas Kurti wanted to introduce in kitchens some tools or techniques that were absent (e.g. vacuum, low temperatures...), and myself, being more of a chemist, I had the idea that any tool from a chemistry laboratory could be helpful for cooking. For example, a fritted glass funnel can clarify stocks in some seconds, decanting bulbs can be helpful for separating fat and stock, and so on. The list of possibilities is tremendous, and indeed almost any chemical tool can be transferred usefully in the kitchen.

Because the application of science is not science, but rather technology, the new way of cooking with modern tools cannot be called molecular gastronomy. In 2000, I proposed that this new way of cooking, using new tools, new ingredients, new methods, should be named "molecular cooking" (or "molecular cuisine" or "molecular cookery"), but the confusion between science and its application was already there, and the whole word was buzzing about molecular gastronomy, often without knowing very well what it was. Even some starred chefs and major food critics wrote that "all ways of cooking are molecular because food is made of molecules!"

Today, as molecular cuisine is already very popular in most countries of the world, and developing in some others, it is important to recall that molecular cooking is not only for fun, but rather a proposal that was introduced because traditional cuisine is not rational enough, in particular because of energy wastes, with a rate of sometimes only 20 %! Also there is no reason why other arts would benefit from the advances of technology, and not cooking.

Some innovations were easy to make. For example, it was obvious to propose some techniques already used in the food industry, such as "alginate pearls" (some people call it spherification, but I prefer the name that I introduced), or using some additives. Indeed one short story about that tells a lot about the way people change their practice: at the beginning of the 80's, when I proposed to chef associations to use such gelling agents as agar, carraghenans, etc., I was told that I would poison people. But when the mad cow disease struck, chefs forgot their remarks and shifted almost immediately toward the use

of such products, fearing gelatin contamination. Today, some even call "gelatins" these gelling agents, ignoring that gelatin is only one of them, with specific properties. In this regard, the book by A. Cazor and Christine Liénard is helpful, because it will tell clearly what can be done in practice and how.

New tools, new ingredients, new methods... One new method was invented by me in 1995, by generalizing whipped cream (called Chantilly cream when sugar is present). I had the idea that if you make an emulsion (like cream) and if you whip it, you would get a foamy emulsion. And this was my invention of "chocolate Chantilly" (but also of "cheese Chantilly," "foie gras Chantilly," etc.). Indeed, this is much more rational than making chocolate mousse, as eggs are useless to produce the product! Rationalization, here, goes along with "home economics," such an important topic that should be taught in all primary schools (and it is now, in France).

Somewhat after the invention of chocolate Chantilly, I proposed a new method for the description of colloidal systems, and this new way of envisioning physical and chemical systems led to various innovations, such as the "eggs at 6X°C" (you cook eggs at low temperature and you get as many results as there are Celsius degrees), or vauquelins, würtz, liebigs, chaptals, braconnots, lavoisiers, faradays... Indeed, for more than ten years, I have been showing one new system (a text describing the "invention" is displayed on the internet site of the chef Pierre Gagnaire), to which I am giving the name of a famous chemist.

In this book, here, some of these innovations are shown, explained, in practice, with recipes that I personally avoid, not being a cook and trying vigorously, as shown in the beginning of this foreword, to avoid the confusion between cooking and science. These recipes are not from me, then, even if many are based on my proposals, but I did nothing in the technical or in the art part of them. The two authors of this book, former students in my Group of Molecular Gastronomy (for a long time the only laboratory entirely devoted to this science), did all alone together, they decided for the tastes, odors, consistencies, associations... and I have the feeling that they did very well. Indeed, this book is very useful because it is simple, and it is true that sometimes long explanations would be boring, and that in the kitchen we do not need scientific knowledge, but rather practical information. Such a book was one of the first on molecular cuisine, and now that siphons, alginate, agar, etc. are in most kitchens of chefs, it is time that a large audience can see how science (molecular gastronomy) could contribute to the advancement of culinary practice.

In the end, Brillat-Savarin was perhaps right to say that the discovery of a new dish makes more for the happiness of humankind than the discovery of a new star. Indeed, it is true that science is a utopia that technology makes alive.

Vive la gourmandise éclairée!

Introduction

Science explores the world, and researches the mechanisms of natural phenomena. Molecular Gastronomy is a scientific discipline studying culinary transformations, as well as phenomena already associated with gastronomy in general. Molecular Gastronomy is part of food science. Technology uses scientific knowledge to produce applications and, in the case of the culinary arts, new applications in the kitchen. Cooking, a mix of art and technique, is inseparable from taste, ingredient quality and craftsmanship. If, in addition, we seek to understand, then cooking becomes molecular cooking. Molecular cooking, or gastronomy, uses applications inspired from technology to create new dishes, new textures, new flavors, and new sensations...

Authors and contributors

ANNE CAZOR is a doctor of Molecular Gastronomy and an engineer for the food industry. She created her own company: *Cuisine Innovation,* a consulting and training firm in France. In this handbook, thanks to her knowledge of technology and science, she selects and explains culinary techniques you will soon easily make your own.

CHRISTINE LIENARD is an engineer in the food industry, holds degrees in food safety and health & nutrition. She has participated in several food research projects and worked in the field of culinary technology. A gastronomically curious person, she offers an eclectic series of gourmet recipes.

JULIEN ATTARD is a free-lance photographer. He has been photo-documenting design elements for the advertising industry.

GUI ALINAT is a chef, instructor and food writer based in Tampa, FL. Classically trained in France, he traveled extensively, working in restaurants across the world. He has become a recognized landmark in the Tampa Bay culinary scene. He is also the author of the award-winning culinary reference *The Chef's Repertoire*.

Ingredients and equipment

As you are about to discover in this book, you will be using texture agents such as gelatin, agar-agar, etc..., and equipment such as precision scale, immersion circulator, etc... Texture agents are used for their specific characteristics: to obtain a heat-resistant gel, for instance, which can't be done with gelatin, but can with agar-agar.

The specific ingredients offered in this book are food additives. A food additive is a substance with or without nutritional value, intentionally added to an ingredient with a specific goal in mind, whether that goal is technological, sanitary, organoleptic or nutritional.

There is no legal, quantitative limit to the consumption of food additives, except for carrageenan (maximum daily dose of 75 mg/kg of body mass). The food additives found in this book should only be used in the proportions described.

The consumption of food additives should be avoided for children under 6 years of age.

Also, some techniques in this book require specific equipment. An oven with an accurate thermostat, or a precision scale are recommended in order to obtain proper results. For example, the use of a precision scale, as opposed to a measuring cup, would give different results. In this type of cuisine, precision is key.

Agar-agar

A gelling agent extracted from red algae.
Widely used in Asian cuisine and produces heat-resistant hard gels
(up to 80°C/176°F).
Recommended dosage: 1% (1 g per 100 g)

Ascorbic acid

An acid obtained by synthesis or extracted from plants.
It is used to acidify preparations and prevent fruits and vegetables from
turning black.

Calcium salt

A source of calcium (lactate and/or calcium gluconolactate).
Increases calcium level as needed for spherification purposes.
Recommended dosage: 1% (1 g per 100 g)

Carrageenan

A gelling agent extracted from red algae.
Obtains heat-resistant, soft, elastic gels (up to 65°C/149°F). Safe daily
dose for human consumption is limited to 75 mg per kg (34 mg per lb)
of body weight.
Recommended dosage: 1% (1 g per 100 g)

Citric acid

An acid obtained by synthesis or extracted from plants.
It is used, for instance, to acidify preparations.

Effervescent sugar

A mix of different sugars into which carbon dioxide (CO_2) has been trapped.

Gelatin

A gelling agent extracted from animal bones. Obtains soft gels that melt when heated.
Recommended dosage: 3% (3 g per 100 g)

Sodium alginate

A thickening/gelling agent extracted from brown algae.
Increases viscosity of low-calcium mixtures and is used, in presence of calcium, for spherification purposes.
Note: Do not pour sodium alginate solutions down the drain as an obstruction is possible (sodium alginate may gel in contact with calcium contained in tap water).
Recommended dosage: 1% (1 g per 100 g)

Sodium bicarbonate

Obtained by synthesis.
It is used, for instance, to acidify preparations.

Soy lecithin

An emulsifier extracted from soy.
It creates emulsions (of water and oil for instance) by stabilizing a liquid dispersing into another.

China cap

A thin-mesh conic strainer. Strains preparations and eliminated useless particles.

Grater

Used to grate or zest ingredients. Must be sharp in order to obtain clean cuts.

Immersion heater

Attached to a bain-marie, an immersion heater heats water to precise and controlled temperatures.

Measuring spoons

Used to measure ingredients (such as food additives) but not as precise as a precision scale. Refer to the conversion table on page 142.

Pipette

Used to take samples of liquid preparations and pour them (drop-by-drop or continuously) elsewhere. Also used as a way to present liquid or semi-solid preparations.

Precision scale

Used to weigh ingredients precisely. We recommend it to weigh food additives.

Slotted spoon

Particularly useful to drain spheres.

Syphon

Creates mousse. Cartridges release NO_2 in the syphon and incorporate into preparations, which aerates them.

Syringe

2 culinary techniques:
Makes droplets for spherification. Also allows the filling of a food-grade tube in order to make gelled "spaghetti".

Tube

Shapes gels into "spaghetti". We recommend food-grade, heat resistant (at least to 100°F/212°F).

Techniques
& applications

Solubilization of sugars

Solubilization consists of dissolving a sugar compound in a liquid. This phenomenon increases with temperature and depends on the quantity and nature of other molecules present in the solution. In the case of sugars, and particularly in the case of saccharose (granulated sugar), one can dissolve approximately 4 lbs per quart of water at 20°C/68°F. Saccharose doesn't dissolve in ethanol or in fat.

50 cl / 2 cups syrup

5 minute prep time

10 minute cook time

1 hour rest time

Ingredients

500 g / 18 oz granulated sugar
50 cl / 2 cups water
50 g / 2 oz dried hibiscus flowers
10 g / 2/3 oz fresh mint

Hibiscus flower and mint syrup

Preparation

- Rinse dried hibiscus flowers
- Bring water and sugar to a boil (boiling is reached once bubbles come up to the surface).
- Take off heat. Add dried hibiscus flowers and mint leaves. Cover and let ingredients infuse at room temperature, until mixture cools off completely (about 1 hour).
- Pass through a sieve, then pour the syrup into a bottle.
- Reserve in refrigerator.
- Dilute 1 dose of syrup with 4 to 5 doses of water to taste.

Sugar is diluted in boiling water. Such a high temperature allows sugar to rapidly dilute in water.

Replace dried hibiscus flowers and mint with any aromatic plant (basil, coriander, lavender, etc...), and let it infuse in hot syrup. This way, you can make surprising homemade syrups for any occasion.

1 chocolate bar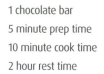

5 minute prep time

10 minute cook time

2 hour rest time

Ingredients

200 g / 7oz dark, white, or milk chocolate
50 g / 2 oz effervescent sugar

Fizzy chocolate

Preparation

∘ Melt chocolate in a double-boiler on low heat. Stir regularly.
∘ When chocolate is melted, take off the double-boiler and whisk vigo-
  rously to emulsify.
∘ Let cool 5 minutes.
∘ Add effervescent sugar in 2 or 3 batches and stir briskly in order to mix
  well.
∘ Pour the preparation onto parchment paper or into silicone molds.
∘ Let it set in a cool, dry place at room temperature for at least 2 hours.

Explore

Effervescent sugar contains carbon dioxide and is mixed with chocolate. Chocolate is rich in cocoa butter (fat), which sugar doesn't solubilize in. When the chocolate cools down, the sugar keeps its effervescent properties because it is protected from humidity contained in the air.

Innovation

Let's add fizz to our cooking! Sprinkle effervescent sugar into your morning coffee, on a pie, or on whipped cream for a sparkly dessert!

Emulsion

An emulsion is a dispersion of two liquids that otherwise do not mix. The dispersed liquid forms droplets in the other. In Cooking, most well-known emulsions are oil and water emulsions. These are not stable emulsions since both liquids end up eventually separating. Nonetheless, tensioactive molecules (soy lecithin, phospholipids, some proteins, such as gelatin, etc...) allow emulsions to stabilize when they position between the dispersed liquid droplets and each other, thus preventing both liquids from separating.

6 servings

20 minute prep time

5 minute cook time

For the Pastis mayonnaise:
1 egg yolk
10 cl / ½ cup vegetable oil
1 tablespoon red wine vinegar
3 teaspoons Pastis
Salt and pepper to taste

For the chicken fries:
600 g / 21 oz boneless,
   skinless chicken breast
2 cups of puffed peanuts
2 eggs
200 g / 7 oz flour
4 tablespoons vegetable oil
Salt and pepper to taste

Puffed peanut chicken fries with Pastis mayonnaise

Preparation

Pastis mayonnaise:
◦ Mix egg yolk and vinegar. Salt and pepper to taste.
◦ Very slowly pour part of the oil to the egg mixture while whisking briskly, in order to emulsify the oil into the egg mixture. When the preparation emulsifies, whisk in the remaining oil.
◦ Whisk in Pastis at the end. Season to taste.

Puffed peanut chicken fries:
◦ Thinly grind puffed peanuts.
◦ Add salt and pepper to taste.
◦ Cut chicken breast into thick "fries."
◦ Flour chicken fries, then add them to the eggs, previously beaten.
◦ Bread them in the ground puffed peanuts.
◦ Deep fry at 350°F for about 5 minutes.
◦ Serve immediately with the Pastis mayonnaise on the side.

Explore

Fat (vegetable oil) is dispersed in water (contained in the egg yolk and vinegar) and forms tiny droplets. Tensioactive molecules (proteins contained in egg yolk) stabilize the emulsion when they position between fat and water droplets. The oil must first be added in small volumes to form an "oil in water" emulsion. If the volume of oil is incorporated too heavily, it will turn into a "water in oil" emulsion and the mayonnaise will not form.

Innovation

Let's replace canola oil with any other liquid fat (olive oil, walnut, clarified butter, etc...); egg yolk with any other ingredient containing tensioactive ingredients (egg white, gelatin, etc...), and let's adjust the quantity of water necessary to add any aromatic ingredient (juice, beer, etc...). Let's work on stunning emulsions such as Light Egg White Mayonnaise, Carrot Aioli, etc...

6 servings

10 minute prep time

3 hour cool time

Ingredients

For the gazpacho:
200 g / 7 oz strawberries
40 g / 1.5 oz pound cake
20 g / ¾ oz pine nuts
6 large leaves of fresh basil
5 cl / 4 tablespoons olive oil
2 cl / 1½ tablespoons cider vinegar
10 cl / ½ cup apple cider

For the strawberry coulis:
100 g / 3.5 oz strawberries
Sugar to taste

Strawberry-pesto-cider gazpacho with strawberry coulis

Preparation

Strawberry-pesto-cider gazpacho:
○ In a stainless steel bowl, roughly mix strawberries which have been cut in half, pound cake, pine nuts, basil leaves, cider vinegar and olive oil. Cover and marinate in the refrigerator for at least 3 hours.
○ Process with hand blender until smooth, then add apple cider to dilute.
○ Finish mixing with hand blender until silky and well-emulsified.

Strawberry coulis:
○ Process strawberries with hand blender until smooth.
○ Add sugar to taste.

To assemble:
○ Fill each shot glasses with gazpacho. Top with strawberry coulis.
○ Serve chilled.

Fat (oil) is dispersed in water (in strawberries, vinegar, cider, etc...) to form tiny droplets. Phospholipids (tensioactive molecules in plant cell membranes) position between oil and water droplets and stabilize the emulsion. The volume of oil, in comparison to water's, is too weak to make a firm "oil in water" emulsion. The thick consistency of gazpacho is due to the added solid elements, such as strawberries, pound cake, etc...

Innovation

A gazpacho is a cold preparation with oil, vinegar, bread, fruits, vegetables, and water, which really is an emulsion. Let's replace olive oil with any other oil (walnut, sesame, etc...); cider vinegar with any other vinegar (balsamic, Jeres, etc...); strawberries with any other fruit or vegetable (beet, cucumber, etc...); cider with another liquid (water, fruit juice, etc...), to create various sweet or savory gazpachos.

Soy lecithin aerated mousse

A mousse is a dispersion of air in a liquid or a solid. In this technique, a mousse is a liquid held together by the tensioactive effect of soy lecithin.

Soy lecithin is a two-part molecule: A part which "like" water (hydrophilic) and another which does not (hydrophobic). These molecules can position between water (which we want to aerate) and air bubbles. This allows for the mousse to stabilize.

The presence of fat will decrease the natural foaming effect of soy lecithin, which will position between water and fat, instead of between water and air bubbles.

50 cl / 2 cups

10 minute prep time

30 minute cool time

15 cl / 5 fl oz dark rum

10 cl / 4 fl oz coffee liquor

25 cl / 1 cup cold coffee

4 g / 1.5 teaspoon soy lecithin

Frothy coffee

Preparation

◦ Mix dark rum, coffee liquor and cold coffee.

◦ Add soy lecithin and blend.

◦ Pour into a syphon and cool in refrigerator for at least 30 minutes.

◦ Add a NO_2 cartridge to the syphon, vigorously shake upside down, and release into a glass from bottom to top.

◦ Serve immediately.

The coffee-based preparation is aerated by the syphon and stabilized by the addition of soy lecithin. Using a syphon incorporates air into the preparation very rapidly. Because of that, soy lecithin can't always stabilize each air bubble in the liquid. Some bubbles disappear while others get stuck at the surface, hence creating a "beer" effect.

Replace coffee by any other low fat liquid. Add soy lecithin (0.3 to 0.8 g for 100 g of preparation), and pour into a syphon. And let's create "beer" with tomato juice, tea, lemonade, etc...

12 maki

6 servings

30 minute prep time

2 minute cook time (optional)

For the mousse:

10 cl / 4 fl oz soy sauce

30 g / 1 oz honey

2 g / ½ teaspoon wasabi

.9 g / ½ teaspoon soy lecithin

For the maki:

300 g / 10 oz sashimi-grade salmon

50 g / 2 oz puffed rice cereal bar

2 nori sheets

2 tablespoon canola oil

Salt

Puffed rice maki and soy sauce foam

For the mousse:

∘ Mix all ingredients and blend vigorously.

∘ Just before serving, use a hand-blender and incorporate as much air as possible into the preparation by frothing the surface.

∘ Reserve the foam. Repeat if necessary.

For the maki:

∘ Slice salmon in order to obtain 2 inch x 1 inch rectangles.

∘ Slice puffed rice cereal bars to obtain rectangles of the same size.

∘ Cut nori sheet in order to obtain 2 inch wide strips.

∘ Place each salmon piece onto a piece of puffed rice cereal bar and wrap in nori.

∘ Optional: Sear maki 30 seconds in hot oil.

∘ Immediately serve the maki with a cloud of soy foam.

Explore

The soy-based preparation is aerated by the hand blender and stabilized with the soy lecithin. The progressive incorporation of air in the preparation (thanks to the action of the hand blender) allows soy lecithin to position between each air bubble and the liquid. Air bubbles are then stuck in the preparation, thus creating a "shampoo" effect.

Innovation

Replace soy sauce by any other low fat flavored liquid. Add soy lecithin (0.3 g to 0.8 g for 100 g of preparation) and hand blend. This way you can create foams with apple juice, garlic, ginger, etc...

This technique brings lightness and softness to spicy or concentrated preparations.

Mousse

A mousse is a foamy emulsion. To obtain a mousse, an emulsion must contain a fat of thick consistency (at cold temperature). Incorporating gas in the preparation can be done by agitation or with a whipped cream dispenser. In both cases, the gas is confined in the emulsion because fat crystallizes when the temperature drops (either on ice or if the gas expands). We then obtain a mousse.

6 servings
10 minute prep time
2 hour cool time

Ingredients

For the raspberry mousse:
20 cl / ¾ cup cream
100 g / 3½ oz seedless raspberry coulis
20 g / 2 tablespoons sugar

For the crumble:
6 plain sugar cookies
18 lychees

Crumble of raspberry and lychee mousse

Preparation

Raspberry mousse:
○ Mix cream, raspberry coulis, and sugar.
○ Pour the preparation into a whipped cream dispenser and refrigerate for 2 hours.

To assemble:
○ Crumble a sugar cookie at the bottom of a glass. Add 3 diced lychees.
○ Insert a gas cartridge into the cartridge node, shake the canister (upside down) and release mousse into the glasses.
○ Serve immediately.

Cream is an emulsion containing water and fat, which is thick when cold. The emulsion is stable thanks to tensioactive agents (milk proteins). The raspberry coulis, which is mostly water, flavors the cream without altering its aeration. A whipped cream dispenser helps obtain a mousse. But the same result can be obtained by chilling and whisking the emulsion on ice.

Let's replace the raspberry coulis with any other aromatic ingredient (spices, Nutella, etc...). Let's mix it with whipping cream in order to obtain a thick emulsion. Then, on ice or with a whipped cream canister, let's add air and crystallize fat with cold. An interesting array of whipped cream will result.

4 servings

20 minute prep time

20 minute cook time

2 hour cool time

Ingredients

For the foie gras foam:
100 g / 3.5 oz foie gras terrine
7 cl / ⅓ cup apple juice

For the boudin créole:
2 large boudin créole
  (125 g / 4.4 oz each)
3 shallots
3 teaspoons duck fat
10 cl / ½ cup apple juice
Salt and pepper to taste

Boudin Créole crème with foie gras foam

Preparation

Foie gras foam:
◦ Temper foie gras terrine at room temperature so it is soft enough to spread.
◦ Mix with apple juice.
◦ Pass preparation through a sieve; it must be very smooth so it doesn't obstruct the whipped cream dispenser.
◦ Pour the preparation into the whipped cream dispenser and refrigerate for 2 hours.

Boudin créole crème:
◦ Mince shallots and salt them.
◦ In a sauté pan, sweat shallots in duck fat on low heat. Take off the heat when they are translucent (about 10 minutes).
◦ In the same pan, sauté boudin créole on each side and poke them with a fork.
◦ Take off the skin.
◦ Use an immersion blender to mix skinless boudin créole, shallots, and apple juice. Salt and pepper to taste.

To assemble:
◦ In each espresso cup, pour the crème.
◦ Add 1 or 2 cartridges into the whipped cream dispenser, heavily shake and top the crème with the foie gras foam.
◦ Serve immediately.

Explore

The foie gras and apple juice preparation is an emulsion containing water (apple juice) and fat (foie gras). The small quantity of tensioactive molecules does not allow for a stable emulsion. Using a whipped cream dispenser is recommended for this foam.

Innovation

Replace foie gras with another solid, fat-based ingredient (chocolate, avocado, blue cheese, etc...) and apple juice with another liquid (orange juice, tea, chicken broth, etc...). Mix both ingredients to obtain a thick emulsion. Then, on ice or with a whipped cream canister, let's add air and crystallize fat with cold.

Soft gel

A gel is a liquid contained in a network. That network can be made up of proteins (such as gelatin, egg proteins, etc...) or polysaccharides (agar-agar, carrageenan, etc...). Gelatin is a protein extracted from meat or fish. This protein is a gelling agent and allows a change of texture from liquid to a gel, thanks to a network forming between the proteins of gelatin. Gelatin dissolves in hot preparations (above 50°C/122°F), and gels at temperatures around 10°C/50°F. If the gel is heated back up to 37°C/98.6°F, it will melt.

If air is added into a preparation while gelling, air bubbles then stabilize (gelatin positions between water and air bubbles) and the preparation gels as it cools. The new gel blocks air bubbles, thanks to the protein network, and we obtain a gelled mousse.

20 minute rest time   **20**

10 servings   **10**

15 minute prep time

30 minute cook time

2 hour cool time

Ingredients

45 cl / 15 fl oz apple juice
15 cl / 5 fl oz water
1 cooked beet (about 125 g / 4½ oz)
30 to 50 g / 1 to 2 oz sugar (to taste)
2 Earl Grey tea bags
10 g / ⅓ oz gelatin (5 sheets)

Apple and beet tea

Preparation

∘ Soak gelatin sheets in cold water to soften.
∘ In a sauce pan, simmer and reduce apple juice by ⅔ to obtain
  15 cl / 5 fl oz of concentrated apple juice. Skim if necessary.
∘ Bring water to a simmer and pour over cooked, diced beet. Let it infuse
  for 2 minutes and strain. Reserve 15 cl / 5 fl oz of beet juice.
∘ In a sauce pan, heat beet juice and concentrated apple juice.
∘ As soon as it simmers, take off the heat and infuse tea bags for 3 or
  4 minutes. Remove tea bags.
∘ Add sugar and reheat. As soon as it simmers again, take off the heat and
  immediately add the softened gelatin sheets. Whisk.
∘ Pour the preparation into an adequate mold, in order to make a 2 to 3 cm
  (about 1 inch) layer. Let it cool at room temperature for 20 minutes, then
  in the refrigerator for 2 hours.
∘ Slice the gel into cubes. Place 2 or 3 cubes in a cup or glass and add
  boiling water.
∘ Mix and enjoy like a tea.

Explore

The preparation contains gelatin, which gels once chilled. Later on, the gel becomes liquid once boiling water is added and releases aromatics. If the "tea" cools down again, it will not gel, since the percentage of gelatin would be too weak to harness the added water.

Innovation

In order to obtain aromatic gels, one can start with any liquid preparation. Bring to a simmer, add gelatin, and chill. What an interesting way to serve a hot beverage!

Dare to work with tea, coffee, herbal infusions, fruit juices, etc... For instance, hot chocolate can be served in a whole new way, with cubed gel of cocoa onto which we could pour hot milk.

4 servings

20 minute prep time

45 minute cook time

2 hour cool time

Ingredients

For the gelled cubes:

3 g / 1 teaspoon / 1.5 sheets of gelatin

50 cl / 2 cups water

1 small marrow bone

1 cinnamon stick

1 star anise

6 coriander seeds

½ onion, diced

1 teaspoon fresh grated ginger

1 tablespoon of sugar

Juice of ½ lemon

2 tablespoon fish sauce

1 pinch salt

For the noodle soup:

100 g / 4 oz rice noodles

200 g / 7 oz beef sirloin

1 onion

2 sprigs cilantro

80 cl / 3 cups water

Pho

Preparation

For the gelled cubes:

○ Soak gelatin sheets in cold water to soften (if using gelatin sheets).

○ In a saucepan, simmer water, marrow bone, cinnamon, star anise, coriander seed, diced onion, ginger, and sugar. Reduce to obtain 8 cl / 3 fl oz of strained liquid.

○ Add lemon juice, fish sauce, and salt.

○ Blend and bring to a simmer.

○ As soon as it simmers, take off the stove and add the gelatin. Whisk to dissolve.

○ Pour the liquid into ice cube trays or into a square container (the gel will then need to be cut into cubes).

○ Let cool at room temperature, then refrigerate for at least 2 hours.

For the noodle soup:

○ Soak the rice noodles in a saucepan filled with tepid water for 30 minutes. Then bring to a boil and strain.

○ In a bowl, plate rice noodles, a few thin slices of raw sirloin, diced onion, and fresh cilantro.

○ Pour 20 cl / 1 cup of boiling water into each bowl and add a gelled cube.

○ Serve immediately.

This recipe explores the properties of gelatin, which melts above 37°C/98°F. Note that gels with no sugar added are slightly less elastic.

Other flavors can be added to a broth or a soup. Add gelatin to a flavored liquid (about 3 g per 100 g of preparation) and let cool. Spice it up with gelled cubes of soy-cilantro-ginger-lime, or aromatic herbs for instance.

Marshmallow

A mousse is a dispersion of gas in a liquid.

If we whip water, the bubbles we create end up gathering on the surface. If we add water into a liquid containing gelatin (tensioactive and gelling molecules), the air bubbles we create are stabilized (gelatin positions between water and air bubbles), and the preparation will gel as it cools. Gelatin activates and traps air bubbles inside a network. We obtain a gelled mousse.

20 pieces

30 minute prep time

8 minute cook time

1 hour rest time

2 hour cool time

Ingredients

300 g / 10 oz sugar + 1 teaspoon
20 cl / ¾ cup mint syrup
2 egg whites
16 g gelatin (8 sheets)
100 g / 4 oz powdered sugar
200 g / 7 oz dark chocolate

Frosty mint and chocolate marshmallow

Preparation

○ Soak gelatin sheets in cold water to soften.
○ In a sauce pan, bring powdered sugar and mint syrup to a boil on moderate heat. Simmer for 7 to 8 minutes. Preparation will foam.
○ Whip egg whites until soft peaks using an electric mixer. Add 1 teaspoon of sugar while whipping.
○ Add gelatin to the mint syrup and whisk to dissolve gelatin.
○ Slowly pour mint syrup into the egg whites and keep on mixing for 5 minutes.
○ Pour a 2 to 3 cm (about 1 inch) layer of preparation into a rectangular mold.
○ Cool at room temperature for 20 minutes, then chill in refrigerator for at least 2 hours.
○ Unmold the marshmallow and dust with powdered sugar, so it's easier to handle.
○ Dice the marshmallow into large square and dust each one with powdered sugar.
○ Melt dark chocolate in a double boiler, on low heat. Stir from time to time. Once the chocolate is thoroughly melted, take off the heat and dip the dices in it (using little tongs or a toothpick), in order to coat them with a thin layer of chocolate.
○ Reserve and let cool. Serve once the chocolate is set.

Beaten egg whites (composed of water and protein) result in a mousse. Adding gelatin to the mousse allows it to gel. We then obtain a gelled mousse.

The gelatin is previously dissolved in the syrup and not in the egg whites, which would present a risk of coagulation of egg white proteins during the heating process. The sweet syrup adds a sticky and elastic texture to the gelled mousse.

Let's replace the mint syrup with any other syrup: lemon, raspberry, caramel, anise... Add it to bloomed gelatin and a meringue and chill for unusual marshmallows.

Ingredients

For the ratatouille:
1 green bell pepper
1 red bell pepper
4 tomatoes
1 large onion
2 garlic cloves
3 tablespoons extra virgin olive oil
Sugar to taste
1 teaspoon paprika
1 teaspoon ground coriander
Salt and pepper to taste

For the gelled mousse:
3 egg yolks
15 cl / 5 fl oz water
½ bouillon cube
  (chicken or vegetable)
6 g gelatin (3 sheets)

Sabayon mousse and cold ratatouille

Preparation

Sabayon mousse:
○ Soak gelatin sheets in cold water to soften.
○ In a sauce pan, bring water and bouillon cube to a simmer.
○ Add gelatin sheets and whisk.
○ Reserve in a large stainless steel bowl and let cool at room temperature (about 30 minutes).
○ Place the bowl on a cold water bath and slowly add the egg yolks while beating with an electric hand mixer, like a sabayon.
○ Quickly transfer the preparation to individual ramekins and cool in the refrigerator for at least 2 hours.

Ratatouille:
○ Small dice onions, tomatoes, red and green peppers. Mince garlic.
○ Sauté onions and peppers in olive oil for 7 to 8 minutes on moderate heat.
○ Add garlic and spices and cook 2 minutes.
○ Add tomatoes, sugar and salt and cook partially covered for about 15 minutes, or until most of the water is evaporated. Season with salt and pepper.
○ Let cool at room temperature, then chill.

To assemble:
○ Unmold sabayon mousse and reserve.
○ Spoon cold ratatouille into a clean ramekin. Top with sabayon mousse and serve.

Explore

Gelatin is dissolved in hot liquid, then egg yolks are added to the preparation. The gelatin is chosen for its tensioactive and gelling properties. On one hand, it allows for blending the liquid with the fat contained in the egg yolks, and, on the other hand, it gels the sabayon.

Innovation

Replace the broth and the egg yolks with any other preparation. Add bloomed gelatin, whisk, and chill. Creations may include gelled mousses of olive oil, carrot juice, tea, thyme, etc...

Coagulation of egg proteins

Egg white contains 90% water and 10% proteins. Egg yolk contains 60% water, 33% fat and 17% proteins.

Some proteins can coagulate (interconnect and form a network) in some conditions (temperature, ph, etc...). Egg white and yolk proteins are naturally different and thus coagulate at different temperatures.

At 61°C/141°F, the first egg proteins coagulate. Then, as the temperature increases, other proteins also coagulate successively, forming a denser and denser network. Depending on the temperature, a wide range of textures can be achieved.

12 pieces 12
15 minute prep time
15 minute cook time

Ingredients

12 grissini
1 large egg
4 black olives, pitted
4 sprigs fresh cilantro
2 tablespoon mayonnaise

Hard-boiled egg lollipop

Preparation

○ Plunge egg into boiling water and cook for 10 minutes. Cool and take shell off.
○ Chop olives. Mince cilantro. Grate hard-boiled egg (large holes).
○ Mix these 3 ingredients.
○ Coat the top part of a grissini with mayonnaise. Roll them in the egg-olive-cilantro mix and serve.

An egg is plunged into water at 100°C/212°F. At this temperature, egg proteins coagulate. The egg becomes hard-boiled with a rubbery white and sandy yolk, which makes it easy to grate.

Egg proteins turn into "networks", which provide great textures to our dishes. Adding whole eggs to a custard makes it thicker. Adding egg yolks (which are high in fat) to a quiche makes it richer. Adding egg white to a flan makes it firmer.

32 pearls

4 servings

15 minute prep time

1 to 2 hour cook time

30 minute rest time

4 egg yolks

15 g / 1 tablespoon powdered sugar

Vanilla-coated, low-temperature egg yolk pearls

In a thermo circulator:

∘ Delicately place raw egg yolks in a freezer zip-lock bag, remove as much air as possible, and immerse into thermo circulator at 67°C/152°F.
∘ Cook for at least 1 hour, then let cool at room temperature.

In a low-temperature oven:

∘ Place whole eggs, directly in their cardboard box, in a low temperature oven at 67°C/152°F. Cook at least 2 hours. Rinse under cold water and let cool at room temperature.

To assemble:

∘ Remove the film around each egg yolk. Separate each yolk in two, then in four in order to obtain 8 similar-sized pieces.
∘ Shape little pearls and roll them in powdered sugar.
∘ Serve immediately.

Explore

Egg yolks are cooked at 67°C/152°F. At that temperature, egg yolk proteins are not all coagulated. That way, the yolks get a creamy, spreadable texture similar to Play-doh.

Innovation

Work those egg yolks into cubes, quenelles, tubes, etc... They can also be used to stuff vegetables for instance.

To obtain different textures, explore a wide range of temperatures (65°C, 66°C, 67°C, 68°C, etc...) to cook egg yolks and egg whites. And pair with sweet or savory preparations.

Meringue

A meringue is three techniques in one: mousse, coagulation of proteins, and heated dehydration. A meringue is achieved with sugar and egg whites whipped firmly, then oven dried.

When egg whites (proteins + water) are whipped, proteins unfold, wrap around water and air bubbles and form a network. We obtain a mousse.

Adding sugar to whipped egg whites increases viscosity, stabilizes the mousse, and makes it firmer.

Baking whipped egg whites dries and cooks them. Egg white proteins coagulate (see coagulation of egg proteins page 56), and water evaporates (see heated dehydration page 86). We obtain a crispy mousse.

8 🥟

◔

●● ●●

3 egg whites
180 g / 6 oz powdered sugar
160 g / 5 oz blue cheese

Blue cheese macaroon

For the macaroons:
- Preheat oven to 90°C/190°F
- Whip egg whites to soft peak.
- Add sugar in small batches while whisking (with an electric blender preferably).
- On a silicone mat, pipe small dollops of meringue.
- Bake for at least 2 hours, with the oven door kept ajar.
- Meringues should not brown, not even slightly, and should be easy to detach from the mat.

To assemble:
- Cut the blue cheese to a diameter similar to the meringues, and .5 cm/ ¼ inch thick.
- Assemble meringues like a sandwich and serve.

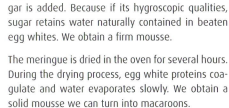

Explore

Egg whites are whipped into a meringue, and sugar is added. Because if its hygroscopic qualities, sugar retains water naturally contained in beaten egg whites. We obtain a firm mousse.

The meringue is dried in the oven for several hours. During the drying process, egg white proteins coagulate and water evaporates slowly. We obtain a solid mousse we can turn into macaroons.

The temperature of 90°C/190°F is chosen to avoid browning of the proteins (see caramelization p. 68 and Maillard reaction p. 74).

Innovation

Meringues can cook at different temperatures to play with different textures: Crunchy (completely dried), or with a soft center (partially dried).

Adjust temperatures to obtain white macaroons (below 95°C/200°F) or golden (higher temperatures).

Also play with different fillings, either sweet or savory.

24 pieces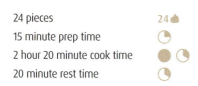

15 minute prep time

2 hour 20 minute cook time

20 minute rest time

Ingredients

4 egg whites

5 cl / 4 tablespoon water

5 mint leaves

5 drops of licorice essence

120 g / 5 oz sugar

120 g / 5 oz powdered sugar

Licorice and mint meringue

Preparation

∘ Preheat oven to 120°C/250°F.

∘ In a small sauce pan, bring water to a simmer. Take off heat and add mint leaves and licorice essence.

∘ Cover with a lid and infuse for at least 20 minutes.

∘ Strain the infusion.

∘ Whip egg whites with an electric mixer. Half way through, add the mint-licorice infusion in small batches, then the sugar, and finally the powdered sugar.

∘ Pipe meringue onto a silicone mat.

∘ Bake 20 minutes. Lower oven temperature to 100°C/210°F and bake another 2 hours.

∘ Take out and cool at room temperature. Serve.

In order to obtain a mousse, we need air, tensioactive molecules, and a liquid. In this recipe, air is incorporated and egg white contains proteins (tensioactive molecules).

If only egg white is used (no added liquid), the volume of the mousse is limited because of its low water content. To create a better mousse, we need to add water during the whipping process.

Explore

After adding sugar, the meringue is dried in a 120°C/250°F oven to accelerate protein coagulation and avoid water loss. Then, it is baked at 100°C/210°F to dry. With that in mind, one can create aerated and crunchy meringues.

Innovation

Replace the licorice-mint infusion by any other flavored liquid such as raspberry or litchi, white or red wine, etc... Let's dry them in the oven to create extremely light meringues that are as light as a cloud.

Caramelization

Caramelization is a non-enzymatic sugar reaction that occurs during a cooking process and which triggers ingredient color and flavor changes.

A caramel occurs when sugar (saccharose) and water are heated to a high temperature (95°C/203°F to 150°C/302°F).

Water then evaporates and saccharose breaks down into two simple sugars: glucose and fructose. As the temperature increases, these simple sugars will reform and create other molecules with deeper colors and aromas.

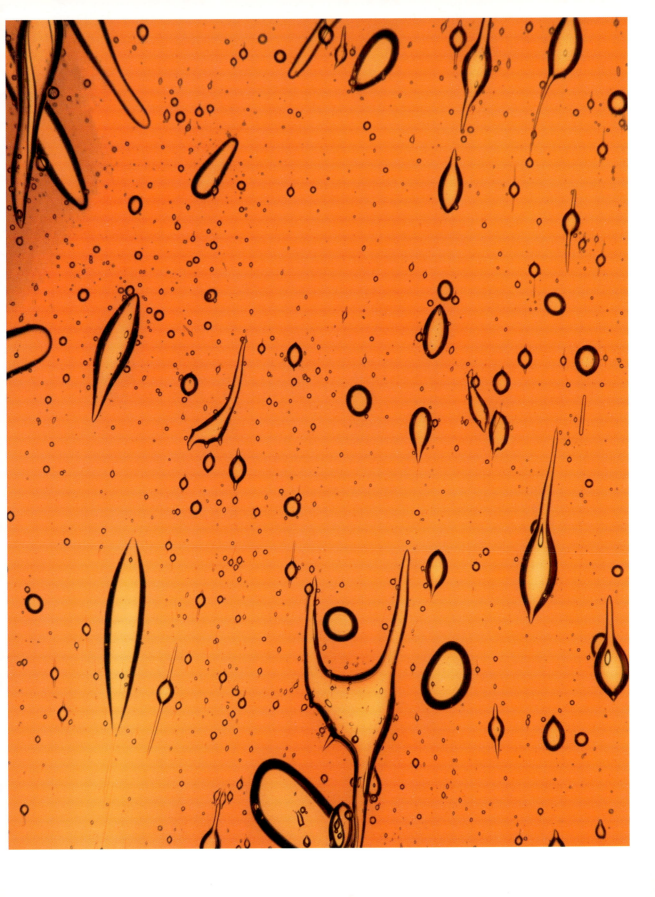

6 servings

10 minute prep time

20 minute cook time

Ingredients

6 large apples (golden delicious)
450 g / 16 oz sugar
6 cl / 2 fl oz water
100 g / 4 oz powdered sugar

Caramel apple

Preparation

- ∘ Wash and take off the stem. Slice off some apple to obtain a cube.
- ∘ Dust with powdered sugar and insert a wooden skewer in each apple.
- ∘ In a saucepan, mix sugar and water and heat to high temperature to make a light caramel (about 15–20 minutes).
- ∘ Take caramel off the heat and dip each apple in, carefully removing excess caramel.
- ∘ Immediately plunge each apple into cold water for 5 seconds. Reserve cubes on a sheet pan lined with parchment paper.
- ∘ Serve immediately.

Note: It is also possible to make bite-sized versions with smaller cubes and toothpicks.

Explore

Sugar is brought to a boil. At such a high temperature, caramelization occurs, which means the sugar
browns and takes on aromas specific to caramel.
Quickly plunging apples into water hardens the caramel without diluting it.

Innovation

Caramel has unique properties. Let's make lollipops
or tuiles. Create contrast between 2 textures (soft
or brittle). Or use it to coat heat-resistant, solid ingredients.

For the mustard caramel:

30 g / 1 oz whole-grain Dijon mustard

3 cl / 1 fl oz beer

100 g / 4 oz sugar

1 teaspoon white vinegar

For the goat cheese flan:

200 g/ 7 oz soft goat cheese

30 cl / 10 fl oz cream

2.5 g / 5/8 teaspoon carrageenan

  (or 1.2 g / ½ teaspoon agar-agar)

Salt and pepper to taste

Goat cheese flan, Dijon mustard caramel

For the mustard caramel:

◦ In a bowl, mix whole-grain Dijon mustard and beer.

◦ In a saucepan, heat up sugar and vinegar at moderate heat, until a light caramel occurs (15–20 minutes).

◦ Take off heat and deglaze with the mustard-beer mix. Stir.

◦ Cook at low heat for 5 additional minutes, in order to get a syrup consistency.

◦ Pour caramel in 6 ramekins.

For the flan:

◦ Mix goat cheese and cream. Salt and pepper to taste.

◦ Heat in a saucepan.

◦ Add agar-agar while whisking. Avoid incorporating too much air.

◦ Bring to a boil for 2 or 3 minutes.

◦ Take off the heat and let cool at room temperature for 5 to 10 minutes.

To assemble:

◦ Pour the custard into the ramekins.

◦ Let cool at room temperature (about 20 minutes) then place in refrigerator (30 minutes).

Sugar is brought to a boil to obtain a caramel. Adding vinegar increases the acidity of the preparation and caramelization is slowed down.

Adding water while deglazing creates a syrupy caramel. If that caramel was cooked longer, water would evaporate and we would get a brittle caramel.

Replace the mustard-beer preparation by any other flavored liquid such as raspberry vinegar, orange juice, etc... When the caramel is slightly brown, deglaze it with that liquid to get a fluid or coating caramel (adjust the quantity of liquid to impact texture). A wide range of caramel can be achieved, such as sweet or sweet and sour, as a side to meat, fruits, vegetables, etc...

Maillard reaction

Another non-enzymatic browning reaction (as opposed to carameli-zation on page 68), a Maillard reaction generates aromatic, flavorful, and colorful molecules. Maillard reactions can be found in bread baking, meat grilling, etc...

A Maillard reaction needs sugars and proteins (or more precisely amino acids) at high temperature. A chain reaction leads to the creation of brown, aromatic, and tasty compounds.

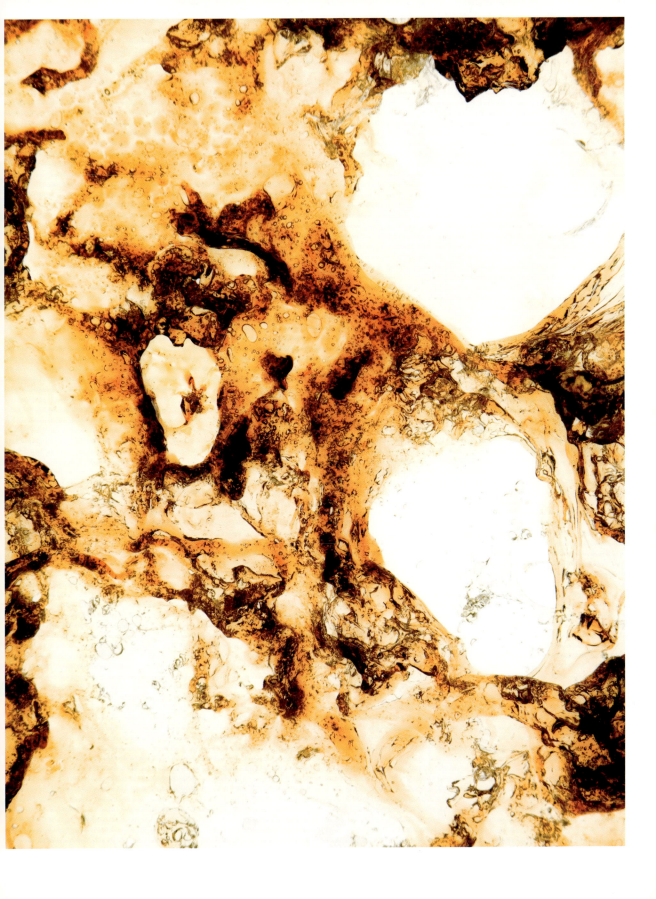

1 can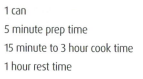

5 minute prep time

15 minute to 3 hour cook time

1 hour rest time

1 can of sweetened condensed milk
   (about 400 g / 14 oz)

Dulce de leche

Preparation

In a pressure-cooker:
○ Put the whole can of sweetened condensed milk into a pressure-cooker.
○ Cover with water and close the cooker.
○ Cook for 15–20 minutes starting from the first pressure.
○ Take off heat and let the pressure-cooker cool completely before opening it (at least 1 hour).

In a saucepan:
○ Put the whole can of sweetened condensed milk in a saucepan.
○ Cover with water and seal with a lid.
○ Boil for 3 hours, adding boiling water regularly so the can is constantly submerged.
○ Take off heat and let completely cool in water before opening it (at least 1 hour).

Sweetened condensed milk contains amino acids and sugar, which are both necessary for a Maillard reaction. High temperatures accelerate Maillard reactions, and the browning and flavoring of dulce de leche occur in the can.

In addition to the Maillard reaction, we can note that caramelization between different sugars is also occurring.

Innovation

Replace sweetened condensed milk with ingredients containing protein and sugar (dairy products, meat, fish, etc...), then heat it up to a high temperature to obtain a Maillard reaction.

6 servings

15 minute prep time

20 minute cook time

20 minute rest time

For the almond cream:
15 cl / 5 fl oz whipping cream
50 g / 2 oz ground almond powder

For the Irish coffee:
30 cl / 10 fl oz whisky
15 cl / 5 fl oz corn syrup
90 cl / 4 cups strong coffee

Irish coffee

For the almond cream:
◦ Preheat oven to 150°C/ 300°F
◦ Spread ground almond powder on a cookie sheet and bake for
  10 minutes. Stir the powder from time to time to avoid burning.
◦ In a saucepan, bring cream with ground almond powder to a simmer.
◦ Take off the heat, put a lid on and infuse for 20 minutes.
◦ Strain the preparation.

For the Irish coffee:
◦ Make coffee
◦ In a saucepan, bring whisky and corn syrup to a simmer. Take off heat.

To assemble:
◦ Pour the sweetened whisky into a glass.
◦ Slowly and delicately pour the hot coffee (use the inside of the glass
  if necessary).
◦ Serve immediately with the almond cream, which can be served in a
  pipette.

Ground almond powder contains amino acid and sugar, which are both necessary for a Maillard reaction. High temperatures accelerate Maillard reactions, and browning and flavoring occurs.

Infusing ground almond powder in the cream brings aromatic and grilled notes, as well as a light brown color.

During production, coffee also goes through a roasting process that brings out its aroma and particular color.

Innovation

Make sure you roast pine nuts, hazelnuts, pistachio, etc... before incorporating into salads, pastries, or beverages, to bring out intense aromas. The roasting effect can supplement your most basic preparations: for instance, pesto with roasted pine nuts or pastries made with various roasted flours.

(Anti)Oxidation

When a fruit or vegetable turns brown when it is cut or bruised, we observe the phenomenon of oxidation. Living cells break and free up both enzymes and phenol molecules, which react once they come in contact with each other. Enzymes modify phenol molecules and produce brown compounds. Ascorbic acid (vitamin C) reduces and prevents oxidation.

4 servings

20 minute prep time

35 minute cook time

For the guacamole:

1 avocado

20 g / 1 oz sugar

Juice of ½ lemon

For the bananas:

2 bananas

Sugar to taste

Guacamole with banana crisp

For the bananas:

∘ Preheat oven to 170°C/340°F

∘ With a knife, slice bananas lengthwise, in order to obtain 6 thin slices (cut the ends if necessary).

∘ Plate the banana slices on a silicone mat and dust with sugar. Bake until dry, about 35 minutes.

∘ Cool at room temperature. Reserve.

For the guacamole:

∘ In a food processor, blend the avocado, sugar, and lemon juice.

To assemble:

∘ Layer bananas and guacamole and serve.

Lemon juice is added to avocado to avoid oxidation. Lemon juice contains ascorbic acid (about 30–40 mg per 100 g), which prevents oxidation. The guacamole thus retains its color longer.

Replace avocado with other fruits or vegetables prone to oxidation (apples, peaches, nectarines, pears, bananas, mushrooms, or artichokes). Mix them with a little lemon juice. You can create an array of fruit or vegetable purees.

6 servings

15 minute prep time

1 hour cool time

Ingredients

1 bottle dry white wine (Sauvignon type)

50 g / 2 oz sugar

1 tablespoon powder sugar

1 teaspoon vanilla extract

2 cl / 1 oz Cointreau

3 drops mint extract

3 g / ½ teaspoon ascorbic acid

2 apples

1 mango

200 g / 7 oz strawberries

20 cl / 1 cup club soda

White sangria

Preparation

○ In a large pitcher, mix white wine, sugar, powdered sugar, vanilla extract, Cointreau, mint extract, and ascorbic acid.

○ Dice fruits, then add them to the white wine mix.

○ Place sangria and club soda in refrigerator for at least 1 hour.

○ Add club soda to sangria and serve.

Explore

Ascorbic acid is added to the sangria as an antioxi-dant to prevent fruits from becoming black. Fruits thus keep their color longer.

Innovation

In preparations prone to oxidation where a lemon taste is not desirable, a small quantity of ascorbic acid (.3 g per 100 g of preparation) can be added. Let's make marinades, seasonings, or smoothies that are capable of withstanding oxidation.

Heated dehydration

Heated dehydration consists of drying ingredients in order to eliminate water by evaporation. Other techniques of dehydration exist (such as salting, cold-smoking, lyophilization, etc...) and are characterized by different water content or drying temperature.

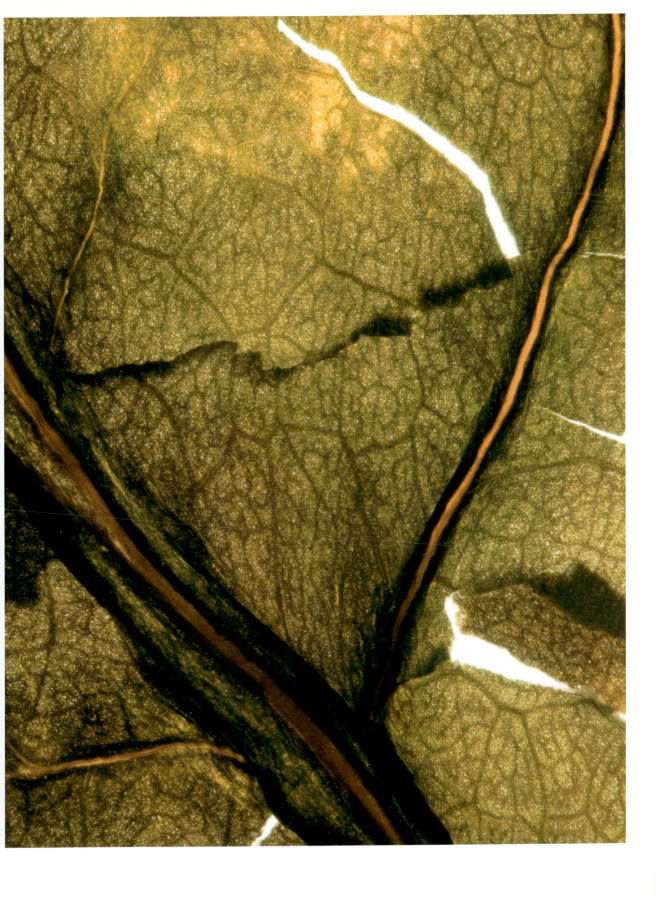

6 servings

15 minute prep time

3 hour cook time

10 marinated black olives

2 garlic cloves

Zest of 3 lemons

10 g / 1 tablespoon of fresh cilantro

Coriander, lemon zest, black olive, and garlic mix

Preparation

- Preheat oven to 100°C/210°F
- Pit olives if necessary and chop them.
- Grate garlic cloves.
- Spread cilantro, grated garlic, lemon zest, and chopped olives in one layer on separate sheet pans.
- Dry in the oven: 20 minutes for the lemon zest, 30 minutes for the garlic, 45 minutes for the cilantro, and 3 hours for the black olives.
- Take stems off the cilantro and add to the other 3 ingredients.
- Serve on grilled meats.

Ingredients are dehydrated thanks to the oven heat: water evaporates. Drying time varies depending on the ingredient, its water content, and its size.

Other aromatics, fruits, vegetables, or charcuterie items can be replaced in this recipe. All kinds of shapes and forms can be achieved: thin slices for fun chips, or chop them finely for a surprising "crumble".

This is a fun technique to personalize dinner parties, but also to conserve fragile or off-season ingredients.

8 pieces

15 minute prep time

2 hour cook time

10 minute rest time

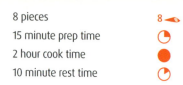

Ingredients

100 g /4 oz canned sweet red peppers
50 g / 2 oz dried apricots
2.5 cl / 1 fl oz water
200 g / 7 oz cottage cheese

Apricot and cream cheese cones

Preparation

◦ Preheat oven to 100°C/210°F.
◦ In a food processor, blend red peppers with the dried apricots and water.
◦ Spread the preparation onto a silicone mat in a very thin layer.
◦ Dry in the oven at least 2 hours. Let cool at room temperature
  (about 10 minutes).
◦ Remove the film and cut into 8 squares. Roll each square into a cone.
◦ Stuff the cones with cottage cheese.
◦ Serve immediately.

Explore

In this recipe, ingredients are blended into a paste. That paste is dehydrated in the oven heat in order to obtain a dry film. Depending on the ingredients used (sugar or starch content, etc...), that film will have different characteristics. Thanks to the sugar content of the dried apricots, the film in this recipe is elastic enough to roll into a cone.

Innovation

Try other fruit or vegetable purees (carrots, peas, prunes, etc...) or starches (purple potato puree for instance). Spread onto a silicone mat and dry in a 100°C/210°F oven to create food-grade films. Make wontons, cones, or any other shape and fill with sweet or savory preparations.

Transfer

A transfer is a compound migrating from one environment to another. A transfer can be generated by osmosis (two different concentrations), by capillarity (physical force), or other ways.

Transfers allow discolorations, colorations, aromatization, etc...

The speed at which compounds transfer is accelerated as temperature increases.

6 servings

5 minute prep time

35 minute cook time

4 minute rest time

Ingredients

For the mint pearls:
40 g / 1 oz tapioca
30 cl / 1 cup water
15 cl / ½ cup mint syrup

For the green tea:
6 green tea bags
2 to 4 tablespoons of sugar

Minty tapioca pearls in green tea

Preparation

For the mint pearls:
◦ In a saucepan, heat up water with mint syrup. Add tapioca to boiling water and cook 30 minutes on low heat, while stirring from time to time.
◦ Take off heat, rinse tapioca under cold water and strain.

For the green tea:
◦ In a glass, place 1 to 2 tablespoons of mint pearls, sugar to taste and 1 green tea bag.
◦ Pour boiling water and infuse 3 to 4 minutes.
◦ Remove green tea bag and serve immediately with a straw.

Explore

Tapioca is a dry ingredient. In presence of a liquid, a migration occurs. In this recipe, the mint syrup contains water, coloring, and flavorings, which all migrate towards the inside of each tapioca pearl. This results in tapioca pearls that are hydrated, green in color, and mint-flavored.

Innovation

Replace tapioca pearls with any other dry ingredients such as rice, couscous, pasta, etc... Also replace mint syrup by any other colorful liquid, such as strawberry syrup, parsley, or beet juice.

1 jar

5 minute prep time

1 hour 15 minute cook time

1 hour 30 minute rest time

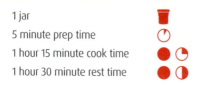

Ingredients

1 apple (granny smith)
100 g / 4 oz red cabbage
40 cl / 2 cups water
Juice of ½ lemon

Sugar (about 1 cup)
1 clove
1 star anise
1 cinnamon stick

Red cabbage and apple gelée

Preparation

○ Place a small plate in the freezer.
○ Wash apple and cut it in batonnets (long, thick sticks). Keep seeds and skin.
○ Cut red cabbage into thin strips.
○ In a saucepan, boil apple, red cabbage and water for 45 minutes.
○ Strain the liquid through a thin mesh and press to extract the maximum of juice.
○ Weigh how much juice was extracted and calculate how much sugar needs to be added (half of the weight).
○ In a saucepan, heat up the juice, sugar, lemon juice, and spices and bring to a boil. Take spices out after 15 minutes and keep cooking for another 15.
○ You know the gelee is ready when a small amount dropped onto the freezing plate ends up gelling instantly.
○ Pour into a jar. Let it cool at room temperature (about 1 hour).
○ Serve with meat, game, or artisan cheese.

Explore

In this recipe, a broth is prepared with water, red cabbage, apple, and spices. The different elements of each ingredient migrate into the broth: red pigments from the cabbage, aromas from the spices, and pectin from the apple. Once it's chilling, pectin gels in contact with sugar. The result is a spicy red gelee. Lemon is added to modify color; acidity naturally modifies the pigments structure.

Innovation

Red cabbage and apple can be replaced by other colorful and tasty ingredients. Let's heat them up in a liquid such as red or white wine, cider, beer, etc...

Broths may be tasted as they are, reduced for sauces, or gelled to accompany meat or fish entrees.

Expansion

Ingredients change volume after they dilate or if some gas is created during the cooking process. Gas can be incorporated into a preparation or come from the transformation of water into steam, for instance. The expansion will be more or less substantial depending on the quantity of gas and how the preparation resists to its escaping.

4 servings

5 minute prep time

20 minute cook time

Ingredients

For the popcorn:
60 g / 2 oz uncooked popcorn
2 tablespoon canola oil

For the caramel sauce:
120 g /5 oz sugar
4 cl / 1 fl oz water
40 g /1 oz salted butter
1 teaspoon cinnamon

Popcorn with caramel sauce

Preparation

For the popcorn:
◦ In a large saucepan, heat up canola oil.
◦ Add corn to the oil and cover.
◦ Let the corn pop while moving the pan from time to time.
◦ Reserve the popcorn.

For the caramel sauce:
◦ In a large saucepan, heat up sugar and water under moderate heat until it gets slightly brown (about 15–20 minutes).
◦ Take off the heat, add cinnamon and diced butter while whisking.

To assemble:
◦ Add popcorn to the caramel sauce in two or three batches. Stir in between batches in order to coat all the grains.
◦ Let popcorn cool at room temperature on a sheet tray.
◦ Break down popcorn if they stick and serve.

Explore

Corn puffs because of the high temperature of the cooking oil. When temperature rises inside each kernel, the small quantity of water evaporates, and starch contained in the grain gels (puffs). Pressure increases and results in a popping effect.

Innovation

Other starchy ingredients can pop. For instance, one can precook, then dry rice, wheat, or quinoa, then heat them up with butter at high temperature in a large saucepan, just like you would corn. This technique opens the door to many creations such as puffed quinoa crumble, crunchy vermicelli nest, etc...

20 pieces

30 minute prep time

5 minute cook time

1 hour rest time

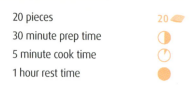

Ingredients

250 g / 8 oz rye flour

12.5 cl / ½ cup tepid water

7 g / 2 teaspoon fresh yeast

3 g / 1 teaspoon salt

Optional: serve with marmalade, spices, olive oil, paté, etc...

Puffed rye bread

Preparation

- Add just enough tepid water to dilute the yeast.
- Mix and rest 10 minutes.
- Put rye flour in a bowl and make a well. Sprinkle with salt on the outside so it is incorporated last.
- Add diluted yeast in the middle of the well, then add tepid water progressively, while mixing from the inside out, to form a dough.
- Knead the dough for 15 minutes. The dough must be soft.
- Make a ball, cover with a clean towel, and let rise for one hour.
- Preheat oven to the maximum setting. Keep a sheet pan inside the oven.
- Roll out the dough thin and cut squares with a chef knife or square cookie cutter.
- Bake the little squares on the hot sheet pan. Once they expanded (about 1 to 2 minutes), let them brown slightly and take out the oven.

era-
npe-

future triggers an expansion of the air brought to the dough during the kneading phase, an expansion of the gas created during fermentation, and the evaporation of water. The gas volume increase and the creation of a gluten network during the kneading phase make the dough expand. When the dough is rolled out thinly, a thin film is created and expands, hence forming a hollow dough.

Explore

Innovation

One of the characteristics of flours is to create networks. Let's use that to make puffed breads of all shapes and forms. Let's use flour, whole wheat flour, cereals, etc...

Keep in mind that many flours do not contain gluten (buckwheat, chestnut, chickpea, corn, etc...).

Spherification

Spherification is the transforming of a liquid into spheres. This technique is achieved by using sodium alginate (a gelling agent extracted from brown algae), which has the property to gel when coming in contact with calcium. These gels are heat-resistant. Basic spherification: sodium alginate is solubilized in a preparation, then that preparation is submerged into a calcium bath. A film instantly gels around the preparation. We obtain a sphere with a liquid center, unstable in time (calcium progresses from the outside in and the sphere gels completely). These spheres must be consumed immediately.

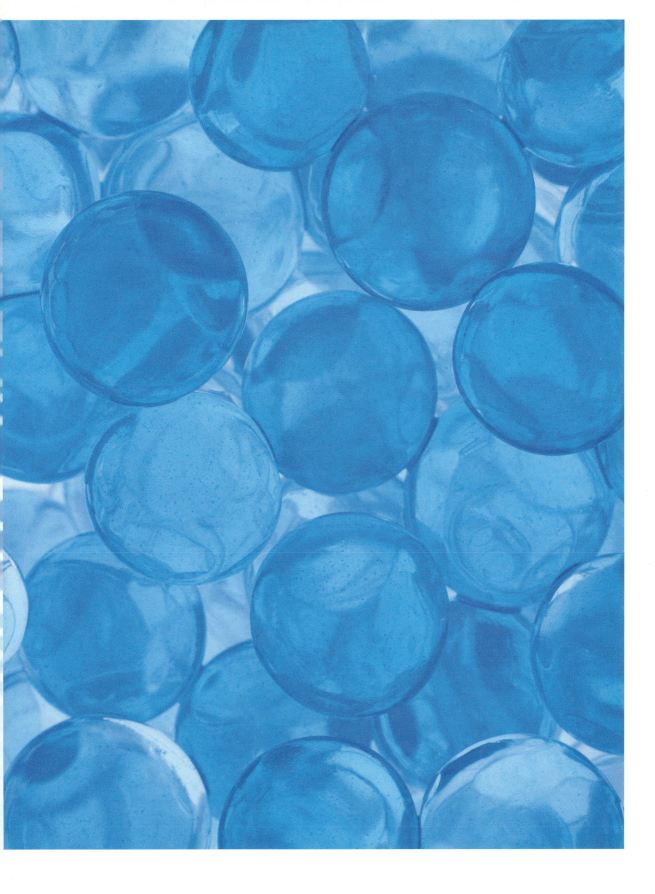

12 spheres   12◉

20 minute prep time ◗

30 minute rest time ◖

2 hour freeze time ❄ ❄

Ingredients

For the calcium bath:
30 cl / 10 fl oz water
3 g / 1 teaspoon calcium salt

For the apple-caramel spheres:
20 cl / ¾ cup apple juice
5 cl / 3½ tablespoon caramel syrup
2.6 g / ½ teaspoon sodium alginate

To assemble:
Vodka

Vodka shot with apple-caramel sphere

Preparation

Calcium bath:
◦ Using an immersion blender, mix calcium salt into water until powder is completely dissolved. Rest 30 minutes.

Apple-caramel spheres:
◦ Mix apple juice and caramel syrup.
◦ Using an immersion blender, mix sodium alginate into apple-caramel mix until powder is completely dissolved. Rest 30 minutes.
◦ Fill half-spherical silicone molds with the preparation and freeze for 2 hours. (This allows for homogenous spheres. This step is optional and a measuring spoon could be used with similar results.)
◦ Plunge the frozen half-spheres into the calcium bath for about a minute. (For better results, the spheres should not touch the sides of the container or float.)
◦ Then plunge them carefully into a plain water bath (use a small, slotted spoon to handle the spheres).

To assemble:
◦ Fill half a shot glass with vodka.
◦ Carefully drop a sphere into each glass.
◦ Serve. Advise your guests to drink the vodka, then pop the sphere into their mouths.

Explore

In this recipe, sodium alginate is dissolved into the apple-caramel mix. When the half spheres are plunged into the calcium bath, a gelled film instantly forms around them and allows the spherification to occur. Even though spheres are rinsed, some calcium stays on the surface and will continue to gel from the outside in. At one point, the sphere will be entirely gelled.

Innovation

The apple-caramel mix may be replaced by any other liquid preparation (fruit juice, herbal tea, vegetable puree, etc...). If sodium alginate is added, the liquid can be turned into a sphere by plunging it into a calcium bath.

Ingredients

For the vinegar pearls:

5 cl / 3 tablespoons raspberry vinegar

10 cl / 7 tablespoons low-calcium
water (below 60 mg/l)

2 cl / 1⅓ tablespoons corn syrup

1.7 g / ½ teaspoon sodium alginate

Red food coloring

For the calcium bath:

30 cl / 10 fl oz water

3 g / 1 teaspoon calcium salt

To assemble:

3 dozen oysters

Raw oyster and its raspberry vinegar pearl

Preparation

Calcium bath:

○ Using an immersion blender, mix calcium salt into water until powder
is completely dissolved. Rest 30 minutes.

Raspberry vinegar pearls:

○ Mix water and corn syrup.

○ Using an immersion blender, mix sodium alginate into preparation
until powder is completely dissolved.

○ Add raspberry vinegar, then red food coloring. Rest 30 minutes.

○ Just before serving, mix the preparation with a whisk.

○ Take a sample with a syringe or a pipette and extract into the calcium
bath drop by drop. After a few seconds, collect the pearls with a slotted
spoon or small strainer, then plunge them carefully in a plain water
bath.

To assemble:

○ Place one or more vinegar pearls into oyster shell and serve
immediately.

Note: Always dispose of sodium alginate in the trash, as opposed to into drains, which would
cause an obstruction.

In this recipe, the preparation is acidic (vinegar-based). Sodium alginate does not dissolve in acidic solutions, unless in two steps. Sodium alginate is dissolved in water first. Then vinegar is progressively added, which slowly reduces the volume of acid in the preparation containing sodium alginate. This two-step technique allows the spherification of acidic preparations.

Play with different liquids. Instead of using strawberry vinegar and corn syrup, use lemon juice, spices, etc... If sodium alginate is added, the liquid can be turned into a sphere by plunging it into a calcium bath.

Reverse spherification

Here, calcium is naturally present in the preparation, and we submerge it into a bath of sodium alginate. A film instantly gels around the preparation. We obtain a sphere with a liquid center, stable in time once it is out of the sodium alginate bath.

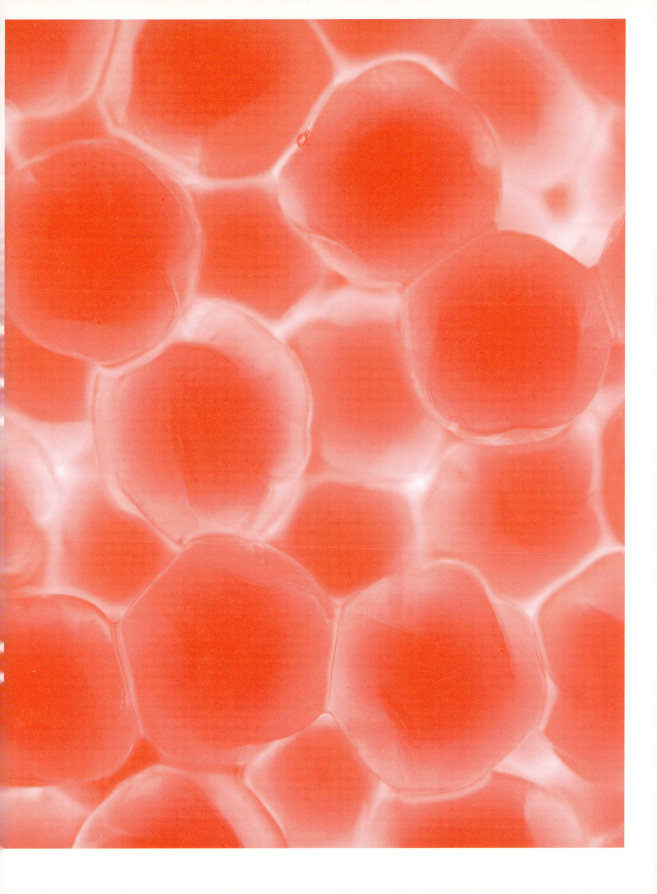

12 spheres

12●

20 minute prep time

45 minute rest time

2 hour freeze time

❄ ❄

Ingrédients

For the sodium alginate:
30 cl / 1⅓ cup low-calcium water
  (below 60 mg/l)
1.5 g / ½ teaspoon sodium alginate

Greek yogurt spheres:
Greek yogurt
1 large garlic clove
Salt and pepper to taste

For the cucumber:
1 cucumber
1 teaspoon of olive oil
1 or 2 teaspoons lemon juice
Coarse sea salt

Spherical tzatziki

Mode Opératoire

Sodium alginate bath:
○ Using an immersion blender, mix sodium alginate into water until powder is completely dissolved. Rest 30 minutes.

Cucumber:
○ Peel and seed cucumber, then grate it or cut very thin slices. Sprinkle with salt and let cucumber release water and drain for at least 15 minutes. Rinse and dry.
○ Add olive oil and lemon juice. Salt and pepper to taste.
○ Plate grated or sliced cucumber in a porcelain spoon. Reserve.

Greek yogurt spheres:
○ Mince garlic. Add to Greek yogurt. Salt and pepper to taste.
○ Fill up the half-spherical silicone molds with Greek yogurt and freeze for at least 2 hours. (This allows for homogenous spheres. This step is optional and a measuring spoon could be used with similar results.)
○ Plunge the frozen half-spheres into the sodium alginate bath for at least 30 seconds. (For better results, the spheres should not touch the sides of the container or float.)
○ Then plunge them carefully in a plain water bath (use a small, slotted spoon to handle the spheres).

To assemble:
○ Delicately place a sphere in each porcelain spoon and let it defrost if necessary.
○ Serve. Inform your guests to taste each preparation in one bite.

Note: Always dispose of sodium alginate in the trash, as opposed to into drains, which would cause an obstruction.

Exploration

Greek yogurt is naturally rich in calcium. When the yogurt preparation is submerged in sodium algi-nate, a film instantly gels around it and allows for spherification. Once the sphere is out of the sodium alginate bath, it stops gelling. Thanks to reversed spherification, the result is a sphere of Greek yogurt with a soft center.

Innovation

Yogurt may be replaced by any other calcium-rich dairy product. It can be turned into a sphere by plunging it into a sodium alginate bath. One can also incorporate spices and aromatic ingredients, grated chocolate, citrus zest, etc... It's a fun way to revisit dairy classics.

12 spheres

20 minute prep time

5 minute cook time

30 minute rest time

2 hour freeze time

Ingredients

For the sodium alginate:
30 cl / 1⅓ cup low-calcium water
(below 60 mg/l)
1.5 g / ½ teaspoon sodium
alginate

For the chorizo spheres:
20 cl / ¾ cup whipping cream
60 g / 2 oz spicy chorizo

1 bottle of cider (alcohol)

Spherical chorizo and cider

Preparation

Sodium alginate bath:
◦ Using an immersion blender, mix sodium alginate into water until powder is completely dissolved. Rest 30 minutes.

Spherical chorizo:
◦ Small dice the chorizo.
◦ In a small saucepan, heat cream and chorizo. As it begins to simmer, take off the heat and stir. Let cool at room temperature for about 20 minutes, then strain.
◦ Fill up the half-spherical silicone molds with the liquid and freeze for 2 hours. (This allows for homogenous spheres. This step is optional and a measuring spoon could be used with similar results.)
◦ Plunge the frozen half-spheres into the sodium alginate for about a minute. (For better results, the spheres should not touch the sides of the container or float.)
◦ Then plunge them carefully into a plain water bath (use a small, slotted spoon to handle the spheres).
◦ Carefully place a sphere on a spoon and let it defrost at room temperature if necessary.

To assemble:
◦ Serve chilled cider in glasses with the spoons of chorizo-flavored cream on the side. Inform your guests to taste it in one bite.

Note: Always dispose of sodium alginate in the trash, as opposed to into drains, which would cause an obstruction.

Explore

Cream is naturally rich in calcium. When a cream-based preparation is plunged into a sodium alginate bath, a film gels around it and allows for spherification. Once the sphere is out of the sodium alginate bath, it will stop gelling. This technique of reversed spherification allows for a sphere of chorizo-flavored cream with a soft center.

Innovation

Replace chorizo with any other strong ingredient (smoked fish, garlic, cocoa, etc...). Infuse it in cream, and let's make a sphere by immersion into a sodium alginate bath.

Brittle gel

Agar-agar is a gelling agent extracted from red algae. Unlike gelatin (a protein), this gelling agent is a polysaccharide (a sugar molecule). It dissolves in hot preparations containing water: Boiling 1 to 3 minutes is appropriate. It gels at about 35°C/95°F. Agar-agar gels are hard and slightly opaque. If it is reheated above 80°C/176°F, it melts.

1 jar serving

10 minute prep time

30 minute cook time

1 hour rest time

30 minute cool time

Ingredients

For the crystal salt:
25 cl / 1 cup water
15 g / ½ oz salt
5 g / 2 teaspoons agar-agar

For the dulce de leche:
1 cup of water
1 7-oz can sweetened,
   condensed milk

Dulce de leche and crystal salt

Preparation

Crystal salt:
◦ In a small sauce pan, bring water and salt to a simmer. Sprinkle agar-agar and mix without incorporating too much air. Simmer 2 minutes. Stir from time to time.
◦ Take off the heat and pour in a mold.
◦ Let cool at room temperature for about 30 minutes, then chill in refrigerator for about 30 minutes.

Dulce de leche:
◦ In a cast iron pan, mix a cup of water with the contents of a 7-oz can of sweetened, condensed milk.
◦ Stir as you cook this mixture, and control the heat so it doesn't bubble over.
◦ Once it turns brown, turn the heat down a little until the mixture reaches your desired shade of color.
◦ Let it cool.

To serve:
◦ Spread dulce de leche on bread.
◦ Shave crystal salt on dulce de leche and enjoy.

Explore

In this recipe, it's the hard-gel properties of agar-agar that are displayed. An agar-agar brittle gel holds up well and is not elastic, which makes it easy to grate. It does not melt in the mouth and has a crunch, unlike a gelatin gel, for instance.

Agar-agar brittle gels are heat-resistant (up to 90°C/176°F) and can be served with hot dishes.

Innovation

Let's bring pizazz to classic recipes! Start with a liquid preparation (fruit or vegetable juice, broth, etc...). Add agar-agar and bring to a boil. Then pour into a mold and chill. Gelled preparation can be added to cold or hot preparations.

4 servings

15 minute prep time

35 minute cook time

10 minute rest time

2 hour cool time

Ingredients

For the honey pearls:

50 g / 3 tablespoons honey

5 cl / 3 tablespoons water

1 g / ⅓ teaspoon agar-agar

50 cl / 2 cups grape seed oil or
   very cold vegetable oil

For the baked camembert:

1 block of camembert

Baked camembert and honey pearls

Preparation

Honey pearls:

○ Place oil in refrigerator a few hours before the preparation.

○ In a small sauce pan, heat water and honey. Sprinkle agar-agar and
   mix without incorporating too much air. Simmer 2 minutes. Stir from
   time to time.

○ Take off heat and cool down at room temperature for 10 minutes.

○ With a syringe or a pipette, take a sample of the preparation and
   plunge it drop-by-drop into the cold oil.

○ Strain and rinse the honey pearls. Reserve.

Baked camembert:

○ Preheat oven to 180°C/350°F.

○ Wrap camembert in aluminum foil.

○ Bake for 30 minutes.

To assemble:

○ Take camembert out of the oven. Make a hole in the center of the
   cheese and add the honey pearls.

○ Serve immediately.

Thanks to principles of physics, when an aqueous preparation is plunged in oil, it does not mix but naturally forms droplets. In this recipe, we explore fast-gelling capabilities of agar-agar. Using oil at a low temperature accelerates the cooling process and we obtain gelled droplets. Note that for very sweet preparations, it is necessary to dilute with enough water to allow the agar-agar to dissolve.

To obtain pearls, let's start with a preparation (coco milk, liquor, lime, etc...). Add agar-agar and bring to a boil. Then pour the liquid drop by drop into a bath of very cold oil. Pearls can be added to cold or hot dishes, and it may be interesting to play with textures: brittle pearls into a soft cheese or a velvety soup, for instance.

Elastic gel

Carrageenan is a gelling agent extracted from red algae. Unlike gelatin (a protein), this gelling agent is a polysaccharide (a sugar molecule). It dissolves in hot preparations containing water above 80°C/176°F; from a practical standpoint, it is appropriate to bring the preparation to a boil. It gels at about 40°C/104°F. Carrageenan gels are elastic and transparent. If a carrageenan is reheated above 65°C/149°F, it melts.

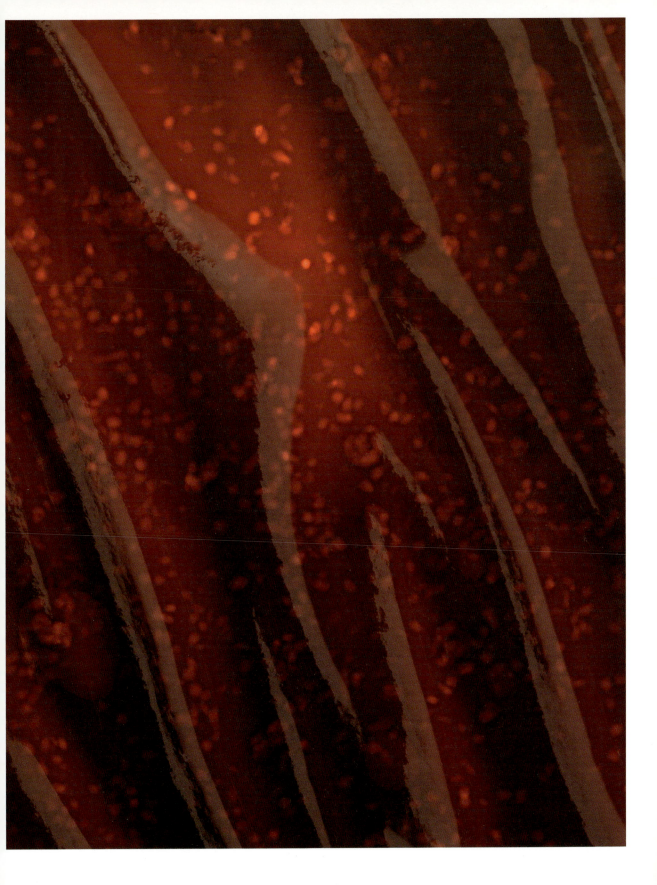

6 servings

15 minute prep time

10 minute cook time

30 minute rest time

For the coco flan:
150 g / 5 oz sweetened condensed milk
30 cl / 10 fl oz coconut milk
30 g / 1 oz grated coconut
5 g / 1¼ teaspoon carrageenan

For the Curaçao spaghetti:
20 cl / ¾ cup Curaçao
4 g / 1 teaspoon carrageenan

Coco flan and Curaçao spaghetti

Coco flan:

○ In a sauce pan, bring coconut milk and sweetened condensed milk to a simmer. Sprinkle carrageenan while whisking, without incorporating too much air.

○ Add grated coconut, take off heat and pour into 6 small ramekins.

○ Let cool at room temperature (about 20 minutes), then refrigerate about 30 minutes.

Curaçao spaghetti:

○ In a small saucepan, heat up the Curaçao. Sprinkle carrageenan while whisking, without incorporating too much air.

○ Connect a syringe (at least 20 ml) to a food-grade silicone tube (about 1 m / 3 ft).

○ As it starts simmering, take saucepan off the heat and plunge the other extremity of the syringe into the Curaçao, so you can suck it into the syringe.

○ Once the tube is filled, plunge it in an ice bath for a few seconds, while maintaining both ends of the syringe above the water line.

○ Push the syringe and extract the Curaçao spaghetti. Reserve.

○ Repeat the process in order to make at least 2 other spaghetti servings. Reheat the sauce to liquefy it if needed.

To assemble:

○ Unmold the coco flans. Garnish them with Curaçao spaghetti and serve.

Note: Without a syringe or a tube, you can make similar tagliatelle if you pour the preparation onto a sheet pan and let cool at room temperature, then cut into ribbons.

Explore

In this recipe, we explore how carrageenan dissolves in alcohol. In spite of the alcohol content, the carrageenan still forms a network and gels the liquor. Elastic (spaghetti) and fast-gelling (flan) properties of carrageenan are also displayed. In addition, note that carrageenan is adequate to gel dairy-based preparations.

Innovation

To quickly make flans or gels without eggs, start with a liquid ingredient (cream, fruit juice, liquor, etc...). Add carrageenan and bring to a simmer. Then pour into a mold and chill. The gel can be cut into "tagliatelle," ribbons, cubes, etc...

6 macaroons

25 minute prep time

20 minute cook time

1 hour 40 minute rest time

For the chocolate génoise:
3 egg whites + 4 egg yolks
40 g / 3 tablespoons sugar + 1 teaspoon
35 g / 1¼ oz brown sugar
75 g / 2⅕ oz flour
35 g / 1¼ oz melted butter
200 g / 7 oz dark chocolate

For the balsamic gelée:
25 cl / 1 cup balsamic vinegar
5 g / 1 teaspoon carrageenan

Chocolate-balsamic macaroon

Chocolate génoise:
∘ Preheat oven to 200°C/400°F.
∘ Layer a sheet pan with parchment paper and grease lightly with butter.
∘ Beat egg yolks with sugar and brown sugar for 5 minutes. Carefully fold in flour, then melted butter.
∘ With a mixer, beat egg whites to a soft peak and fold them into the egg yolks.
∘ Pour onto the sheet pan (1 cm / ½ inch thick). Bake 10 minutes.
∘ Turn génoise over on another sheet pan. Cover with a clean towel and let cool at room temperature for about 30 minutes.
∘ Melt dark chocolate in double boiler.
∘ Cut the génoise into 5 cm / 2 inch circles with a cookie cutter.
∘ Dip circles into the chocolate and let rest for 1 hour, or until chocolate has set.

Balsamic vinegar gelée:
∘ In a small saucepan, bring the vinegar to a simmer. Sprinkle carrageenan while whisking, without incorporating too much air.
∘ When it begins to simmer, take off the heat and pour into a mold (0.5 cm / ⅓ inch thick).
∘ Let cool at room temperature for about 20 minutes.

To assemble:
∘ Cut circles into the gelée with a cookie cutter (0.5 cm / ⅓ inch thick).
∘ Make macaroons with 2 circles of génoise and a circle of gélee in between. Serve.

Explore

In this recipe, we explore how carrageenan gels in acidic solutions. Vinegar is acidic (low pH), but carrageenan still forms a network and gels. It is also possible to use agar-agar in acidic preparations.

Innovation

Gelled ingredients can be added to classic dishes. Start with an acidic liquid (cider vinegar, citrus juice, etc...). Add carrageenan and bring to a simmer. Then pour into a mold and chill. The gel can be cut into "tagliatelle," ribbons, cubes, etc... and incorporated into salads, crepes, cakes, etc...

Effervescence

Effervescence is the creation of gas bubbles in a liquid. It occurs when an acid comes in contact with soda bicarbonate. For instance, if we mix soda bicarbonate with citric acid and add water, a chemical reaction occurs and releases carbon dioxide.

6 servings

10 minute prep time

13 minute cook time

30 minute rest time

2 hour 30 minute cool time

For the beer foam:
20 cl / ¾ cup beer
10 g / 3 tablespoons sugar
4 g gelatin
2 g sodium bicarbonate

For the lemon gelée:
Juice of 2 lemons
  (about 10 cl / 1/3 cup)
50 g / 2 oz sugar
3 g / 1 teaspoon agar agar

Beer foam and lemon gelée

For the beer foam:
○ Soak gelatin sheets in cold water to soften them.
○ In a sauce pan, bring beer and sugar to a simmer, then take off heat and add gelatin while whisking.
○ Let cool at room temperature (about 10 minutes).
○ Add sodium bicarbonate while whisking.
○ Pour preparation into a canister.
○ Cool in refrigerator for 2 hours.

For the lemon gelée:
○ In a saucepan, heat up lemon juice and sugar. Sprinkle in agar-agar and whisk without incorporating too much air. Simmer for 2 to 3 minutes while stirring.
○ Take off heat and pour preparation into 6 small glasses.
○ Let cool at room temperature (about 30 minutes), then cool in refrigerator (about 30 minutes).

To assemble:
○ Add a cartridge in the canister and shake vigorously.
○ Release in small glasses (the volume of beer foam must be twice as much as the volume of lemon gelée in order to get an optimal fizzy taste).
○ Serve immediately.

Explore

Citric, ascorbic acid (naturally present in lemon juice), and sodium bicarbonate are presented in two different textures in this recipe. Lemon juice is gelled and sodium bicarbonate is solubilized in the beer foam. When tasted, both elements come in contact. Carbon dioxide is freed and creates a fizzy taste.

Innovation

Replace lemon gelée by other acidic preparations (balsamic vinegar gelée, citrus mousse, etc...) and mix it with mango mousse or soda gelée, to which we can add sodium bicarbonate (up to 1 g for 100 g). Mixing these 2 elements will bring new exciting sensations.

6 servings

5 minute prep time

20 minute cook time

30 minute rest time

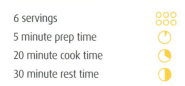

Ingredients

100 g / 3 oz sugar
5 cl / ¼ cup water
.2 g citric acid
.2 g sodium bicarbonate

Fizzy caramel lollipop

Preparation

○ Dilute sodium bicarbonate into 1 tablespoon of water.
○ In a saucepan, bring sugar, water, and citric acid to a simmer (do not stir) until a light brown caramel occurs, about 15–20 minutes.
○ Remove from heat and add the sodium bicarbonate while whisking vigorously.
○ Pour 6 tiny dollops of caramel onto a silicone mat. Place a toothpick on each dollop sideways to form a lollipop.
○ Let the caramel harden at room temperature (about 30 minutes).
○ Reserve in refrigerator.

Sugar, citric acid, and water are heated up. When sodium bicarbonate is added, it reacts with citric acid. Carbon dioxide is released, which turns the preparation into foam.

In addition, sodium bicarbonate tends to decrease acidity, which promotes caramelization. When it cools down, the caramel traps bubbles of carbon dioxide.

Innovation

Spark up your recipes by associating an acid and a base into an aqueous preparation. Let's trap air bubbles in preparations that set rapidly (such as caramel). A lemonade could be presented to guests and become fizzy with the tableside addition of sodium bicarbonate.

Fermentation

Fermentation is due to micro-organisms developing in presence of water and nutritious substances, at proper temperatures.

We distinguish between 2 types of fermentation: Alcoholic (beer, bread, etc...), which transforms sugars into alcohol and carbon dioxide (CO_2), and lactic fermentation (yogurt, cheese), which transforms sugars into lactic acid.

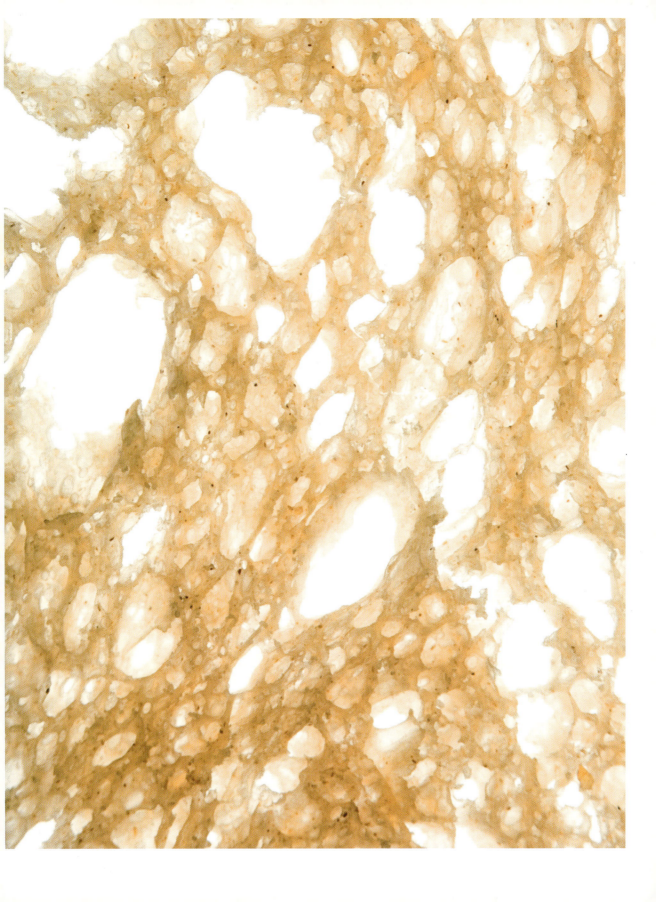

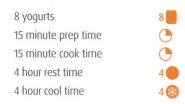

8 yogurts

15 minute prep time

15 minute cook time

4 hour rest time

4 hour cool time

8

Ingredients

For the carrot syrup:
10 cl / 1/3 cup carrot juice
Juice of ¼ lemon
50 g / 2 oz sugar

For the yogurt:
1 plain yogurt
1 qt whole milk
4 tablespoon sugar
Cilantro to taste

Cilantro-infused yogurt and carrot coulis

Preparation

For the carrot syrup:
- In a saucepan, bring to a simmer carrot juice, lemon juice, and sugar, and boil until it reduces by half (about 10 minutes).
- Transfer the syrup to a bowl and let it cool at room temperature (carrot juice should have the consistency of honey after cooling down).

For the yogurt:
- In a saucepan, bring to a simmer half of the milk, sugar, and coriander while stirring from time to time. Take off heat and strain.
- Mix the other half portion of milk with the yogurt.
- Pour hot milk on the cold milk/yogurt while stirring vigorously. The preparation should now be around 40°C/104°F.
- Place 1 tablespoon of carrot syrup at the bottom of each jar, then delicately pour the yogurt preparation on top.
- Place the jars in a low temp oven set at 40°C/104°F.
- Let the yogurt set for about 4 hours. Cool in refrigerator for 4 hours.

Yogurt is a natural ferment and brings microorganisms (lactobacillus bulgaricus and streptococcus thermophilus) that are necessary to lactic fermentation.

Water (contained in milk), sugar (added or already contained in milk), and temperature create a perfect environment for those microorganisms to develop.

During lactic fermentation, microorganisms produce lactic acid, which modifies the structure of milk proteins and leads to their coagulation: yogurt.

Replace a plain yogurt by another source of microorganism such as kefir, for instance. Mix it with milk and ferment during several hours (depending on ferment) at 40°C/104°F. The results can be homemade fermented milks of various consistency and more or less acidic. An essence can then be added: almonds, mint, spices, etc...

6 servings

5 minute prep time

3 minute cook time

1 hour rest time

Overnight refrigeration

10 ❄

Ingredients

1 qt grape juice
2.5 g fresh yeast

Fermented grape juice

Preparation

○ Mix fresh yeast with just enough tepid grape juice to dilute.
○ Mix and rest 10 minutes.
○ In a saucepan, bring the rest of the grape juice to 40°C/104°F.
○ Add the yeast to the grape juice.
○ Mix and cover with clean towel. Ferment for about 1 hour in a warm place (in a preheated oven at 50°C/122°F that has been turned off, for instance).
○ Transfer to refrigerator overnight.
○ Serve chilled.

Fresh yeast is a ferment and brings the microorganisms necessary to the alcoholic fermentation. Water and sugar (contained in the grape juice), as well as the temperature (40°C/104°F) create a perfect environment for microorganisms to develop. During an alcoholic fermentation, microorganisms produce carbon dioxide, which carbonate the grape juice.

Replace grape juice with another juice (apple juice, cranberry juice, etc...). Add a little fresh yeast (.2 g to .3 g per 100 g) and ferment for a few hours in a warm place. When chilled, the fermented drink is refreshing and fizzy to the taste. A nice way to create alternative drinks.

Appendices

Weight/dosage conversion

Each spoon must be used with the scoop-and-sweep method (fill and flush with a knife).

This conversion table is only given for your information. Powders have different characteristics (the granulometry is different, for instance) depending on suppliers. Therefore, we recommend using a precision scale (0.1 g increments).

Make sure you wash and dry spoons between the ingredients.

Example of calculation:

◦ If 0.8% means 0.8 g of powder for 100 g of preparation, how much agar-agar should we use for 300 g of preparation?

 0.8 g of agar-agar corresponds to 100 g of preparation.

◦ If x g of agar-agar corresponds to 300 g of preparation, we obtain:
 x/0.8 = 300/100. So x = (300/100) X 0.8 = 2.4 g.

◦ If we refer to the following table conversion, one can measure 2.4 g of agar-agar using:
 One 0.63 ml spoon + one 1.25 ml spoon + one 2.5 ml spoon.

Grams/volume conversion table

(en g)	0.63 ml 1/8 teaspoon	1.25 ml 1/4 teaspoon	2.5 ml 1/2 teaspoon	5 ml 1 teaspoon	15 ml 1 tablespoon
Agar agar	0.4	0.6	1.4	2.8	8.3
Sodium alginate	0.6	0.8	1.9	3.9	11.6
Calcium salt	0.5	0.7	1.5	3.2	9.5
Carrageenan	0.6	0.9	2	4.1	12.1
Citric acid	0.6	0.8	1.9	3.9	12.1
Sodium bicarbonate	0.8	1.2	2.6	5.5	16.5
Ascorbic acid	0.7	1.0	2.3	4.8	14.0
Soy lecithin	0.4	0.5	1.2	2.6	8.1

Glossary

Aqueous
Relating to water, or a water solution.

Aroma
Pleasant smell of a plant, which could be described as "aromatic plant."

Coagulation
A protein association that forms a network. Usually triggered by physical agents such as heat, agitation, etc... or chemicals such as ph, enzymes, etc...

Dispersion
Solid, liquid, or gas containing another body that is uniformly spread out into it.

Dissolution
From "dissolving." A homogenous mix of reactionless substances.

Enzyme
A protein that accelerates chemical reactions. Enzymes are biological catalysts.

Ferment
Micro-organisms triggering fermentation.

Hygroscopic
The property of a compound to absorb water present in the air.

Micro-organism
A microscopic, living being such as yeast or bacteria.

Miscibility
The property of a compound to mix into a liquid.

Molecule
atoms united by more or less stable chemical ties.

Network
A group of large molecules linked together.

Organoleptic

The cognitive property of a food that has an effect on senses, whether visual, olfactive, tactile, gustative, or auditive).

Ph

A measure of acidity and alkalinity of a solution that is a number on a scale. Acidic solutions have a ph inferior to 7, alkaline solutions are superior to 7, and neutral are equal to 7.

Polysaccharides

A long chain of sugar molecules.

Proteins

A long chain of amino acids.

Solubilization

A synonym of dissolution (see page 146).

Taste

A global and synthetic sensation that one has while eating.

Tensioactive

A tensioactive material is one that changes the value of surface tension. Those molecules are composed with a water soluble and insoluble part, which can be used as an emulsifier.

Viscosity

The resistance of a fluid to pour.

Suppliers of ingredients for thickening, gelling and emulsification:*

North America

Culinary Imports
141 Newburyport Tnpk.
Suite 194
Rowley, MA 01969
Phone: (978) 621-9494
Fax: (978) 462-0398
email: info@culinary-imports.com
http://www.culinary-imports.com

La Tienda
1325 Jamestown Rd
Williamsburg, VA 23185Orders: 800 710-4304
Customer Service: 888 331-4362
http://www.tienda.com

Le Sanctuaire
315 Sutter Street, 5th Floor
San Francisco, CA 94108
T: 415-986-4216
F: 415-986-4217
info@le-sanctuaire.com
http://le-sanctuaire.com

L'Epicerie
Attn: Customer Service
106 Ferris Street
Brooklyn, NY 11231
Toll free: 866-350-7575
Local: 718-596-7575
Fax: 718-596-6444
customer_service@lepicerie.com
http://www.lepicerie.com

Paris Gourmet
145 Grand Street
Carlstadt, NJ 07072
201-939-5656
http://parisgourmet.com

Terra Spice Company
P.O. Box 26
605 Roosevelt Road
Walkerton, Indiana 46574
574-586-2600
info@terraspice.com
http://www.terraspice.com

TIC Gums, Inc.
10552 Philadelphia Rd
White Marsh, MD 21162
Phone (410) 273-7300
Fax (410) 273-6469
info@ticgums.com
http://www.ticgums.com

WillPowder
Ciao Imports
1521 Alton Road Suite #325 .
Miami Beach, FL 33139
Phone: 866.249.0400
Fax: 866.353.8866
Email: spice@willpowder.com
http://www.willpowder.net

Saveurs Molécule-R inc.
5425 rue de Bordeaux, suite 402
Montréal (Qc), Canada
H2H 2P9
TEL.: (514) 564-3363
http://www.molecule-r.com

Europe

Cream Supplies
Catering & Leisure Supplies Ltd
Unit B5, Mountbatten Bus Park,
Jackson Close,
Farlington, Portsmouth,
PO6 1US.
UK Tel: 0845 226 3024.
International Tel: +44 1243 200401
Fax: 0845 226 3014
E-Mail: sales@creamsupplies.co.uk
http://www.creamsupplies.co.uk

Cuisine Innovation
2 rue Claude Bernard
21000
Dijon
France
E-mail: contact@cuisine-innovation.fr This e-mail address is being
             protected from spam bots, you need JavaScript enabled to
             view it
Telephone: 09 52 13 78 69
Fax: 09 57 13 78 69
http://www.cuisine-innovation.fr

Infusions4chefs Ltd
4 Lundy Court
Rougham Industrial Estate
Rougham
Bury St Edmunds
Suffolk IP30 9ND
Tel: 01359 272577
Fax: 01359 272588
info@infusions4chefs.co.uk
http://www.infusions4chefs.co.uk/page/contact.html

Australia

The Melbourne Food Ingredient Depot
508 Lygon Street
East Brunswick,
Victoria, Australia
Phone 03 9386 3206
(International Phone: + (61 3) 9386 3206
Facsimile: (+61 3) 9386 1888
http://www.mfcd.net

The Red Spoon Company
461 Liverpool Rd
Strathfield New South Wales 2135
Australia
Ph: 02 9758 7459, Fax: 02 9642 4821.
http://www.redspooncompany.com

*This list was prepared with the assistance of Martin Lersch and his
Khymos blog, which can be found at http://blog.khymos.org.

Indices

Index
of ingredients

Caramel: Caramel apple 60; Dulce de leche and crystal salt 108; Fizzy caramel lollipop 122; Goat cheese flan, Dijon mustard caramel 62

Carrageenan: Chocolate-balsamic macaroon 116; Coco flan and Curaçao spaghetti 114

Carrot juice: Cilantro-infused yogurt and carrot coulis 126

Caramel syrup: Vodka shot with apple-caramel sphere 96

Chicken: Puffed peanut chicken fries with Pastis mayonnaise 18

Chocolate, dark: Chocolate-balsamic macaroon 116; Fizzy chocolate 14; Frosty mint and chocolate marshmallow 42

Chocolate, milk: Fizzy chocolate 14

Chocolate, white: Fizzy chocolate 14

Chorizo: Spherical chorizo and cider 104

Cider, hard: Spherical chorizo and cider 104

Cider vinegar: Strawberry-pesto-cider gazpacho with strawberry coulis 20

Cilantro, fresh: Cilantro-infused yogurt and carrot coulis 126; Coriander, lemon zest, black olive, and garlic mix 78; Hard-boiled egg lollipop 48; Pho 38

Cinnamon, ground: Popcorn with caramel sauce 90

Cinnamon stick: Pho 49; Red cabbage and apple gelée 86

Citric acid: Fizzy caramel lollipop 122

Cloves: Red cabbage and apple gelée 86

Club soda: White sangria 74

Coconut, grated: Coco flan and Curaçao spaghetti 114

Coconut milk: Coco flan and Curaçao spaghetti 114

Coffee: Frothy coffee 24; Irish coffee 68

Coffee liquor: Frothy coffee 24

Cointreau: White sangria 74

Condensed milk, sweetened: Coco flan and Curaçao spaghetti 114; Dulce de leche 66

Coriander, ground: Sabayon mousse and cold ratatouille 44

Coriander seeds: Pho 38

Corn syrup: Irish coffee 68; Raw oyster and its raspberry vinegar pearl 98

Cottage cheese: Apricot and cream cheese cones 80; Goat cheese flan, Dijon mustard caramel 62

Crème: Crumble of raspberry and lychee mousse 30; Dulce de leche and crystal salt 108; Goat cheese flan, Dijon mustard caramel 62; Irish coffee 68

Cucumber: Spherical tzatziki 102

Curaçao: Coco flan and Curaçao spaghetti 114

Dark rum: Frothy coffee 24

Dijon mustard: Goat cheese flan, Dijon mustard caramel 62

Duck fat: Boudin Créole crème with foie gras foam 32

Earl Grey tea: Apple and beet tea 36

Effervescent sugar: Fizzy chocolate 14

Eggs: Hard-boiled egg lollipop 48; Puffed peanut chicken fries with Pastis mayonnaise 18

Egg whites: Blue cheese macaroon 54; Chocolate-balsamic macaroon 116; Frosty mint and chocolate marshmallow 42; Licorice and mint meringue 56

Egg yolks: Chocolate-balsamic macaroon 121; Puffed peanut chicken fries with Pastis mayonnaise 18; Sabayon mousse and cold ratatouille 44; Vanilla-coated, low-temperature egg yolk pearls 50

Fish sauce: Pho 38

Flour: Chocolate-balsamic macaroon 116; Puffed peanut chicken fries with Pastis mayonnaise 18

Foie gras: Boudin Créole crème with foie gras foam 32

Garlic: Coriander, lemon zest, black olive, and garlic mix 78; Sabayon mousse and cold ratatouille 44; Spherical tzatziki 102

Gelatin: Apple and beet tea 36; Beer foam and lemon gelée 120; Frosty mint and chocolate marshmallow 42; Pho 38; Sabayon mousse and cold ratatouille 44

Ginger, fresh: Pho 38

Grape juice: Fermented grape juice 128

Grape seed oil: Baked camembert and honey pearls 110

Greek yogurt: Spherical tzatziki 102

Green pepper: Sabayon mousse and cold ratatouille 44

Green tea: Minty tapioca pearls in green tea 84

Grissini: Hard-boiled egg lollipop 48

Hibiscus flowers, dried: Hibiscus flower and mint syrup 12

Honey: Baked camembert and honey pearls 110; Puffed rice maki and soy sauce foam 26

Lemon juice: Beer foam and lemon gelée 120; Cilantro-infused yogurt and carrot coulis 126; Guacamole with banana crisp 72; Pho 38; Red cabbage and apple gelée 86; Spherical tzatziki 102

Lemon zest: Coriander, lemon zest, black olive, and garlic mix 78

Licorice essence: Licorice and mint meringue 56

Low-calcium water: Raw oyster and its raspberry vinegar pearl 98; Spherical chorizo and cider 104; Spherical tzatziki 102

Lychees: Crumble of raspberry and lychee mousse 30

Mango: White sangria 74

Marrow bone: Pho 38

Mayonnaise: Hard-boiled egg lollipop 48; Puffed peanut chicken fries with Pastis mayonnaise 18

Milk, whole: Cilantro-infused yogurt and carrot coulis 126

Mint, fresh: Hibiscus flower and mint syrup 12; Licorice and mint meringue 56

Mint extract: White sangria 74

Mint syrup: Frosty mint and chocolate marshmallow 42; Minty tapioca pearls in green tea 84

Nori: Puffed rice maki and soy sauce foam 26

Olive oil: Sabayon mousse and cold ratatouille 44; Spherical tzatziki 102; Strawberry-pesto-cider gazpacho with strawberry coulis 20

Onion: Pho 48; Sabayon mousse and cold ratatouille 44

Oysters: Raw oyster and its raspberry vinegar pearl 98

Paprika: Sabayon mousse and cold ratatouille 44

Pastis: Puffed peanut chicken fries with Pastis mayonnaise 18

Pine nuts: Strawberry-pesto-cider gazpacho with strawberry coulis 20

Popcorn, uncooked: Popcorn with caramel sauce 90

Pound cake: Strawberry-pesto-cider gazpacho with strawberry coulis 20

Powdered sugar: Caramel apple 60; Frosty mint and chocolate marshmallow 42; Licorice and mint meringue 56

Puffed peanuts: Puffed peanut chicken fries with Pastis mayonnaise 18

Puffed rice cereal bar: Puffed rice maki and soy sauce foam 26

Raspberry coulis: Crumble of raspberry and lychee mousse 30

Raspberry vinegar: Raw oyster and its raspberry vinegar pearl 98

Red cabbage: Red cabbage and apple gelée 86

Red food coloring: Raw oyster and its raspberry vinegar pearl 98

Red pepper: Sabayon mousse and cold ratatouille 44

Red wine vinegar: Puffed peanut chicken fries with Pastis mayonnaise 18

Rice noodles: Pho 38

Rye flour: Puffed rye bread 92

Salmon, sashimi grade: Puffed rice maki and soy sauce foam 26

Salt: Boudin Créole crème with foie gras foam 32; Goat cheese flan, Dijon mustard caramel 62; Pho 38; Puffed peanut chicken fries with Pastis mayonnaise 18; Puffed rice maki and soy sauce foam 26; Puffed rye bread 92; Sabayon mousse and cold ratatouille 44; Spherical tzatziki 102

Shallots: Boudin Créole crème with foie gras foam 32

Sodium alginate: Raw oyster and its raspberry vinegar pearl 98; Spherical chorizo and cider 104; Spherical tzatziki 102; Vodka shot with apple-caramel sphere 96

Sodium bicarbonate: Beer foam and lemon gelée 120; Fizzy caramel lollipop 122

Star anise: Red cabbage and apple gelée 86; White sangria 74

Strawberries: Strawberry-pesto-cider gazpacho with strawberry coulis 20; White sangria 74

Soy lecithin: Frothy coffee 24; Puffed rice maki and soy sauce foam 26

Soy sauce: Puffed rice maki and soy sauce foam 26

Sugar cookies: Crumble of raspberry and lychee mousse 30

Sugar: Apple and beet tea 36; Beer foam and lemon gelée 120; Blue cheese macaroon 54; Caramel apple 60; Chocolate-balsamic macaroon 116; Cilantro-infused yogurt and carrot coulis 126; Fizzy caramel lollipop 122; Goat cheese flan, Dijon mustard caramel 62; Guacamole with banana crisp 72; Hibiscus flower and mint syrup 12; Licorice and mint meringue 56; Minty tapioca pearls in green tea 84; Pho 38; Popcorn with caramel sauce 90; Red cabbage and apple gelée 86; Strawberry-pesto-cider gazpacho with strawberry coulis 20; White sangria 74

Sweet red peppers, canned: Apricot and cream cheese cones 80

Tapioca: Minty tapioca pearls in green tea 84

Tomatoes: Sabayon mousse and cold ratatouille 44

Vodka: Vodka shot with apple-caramel sphere 96

Wasabi: Puffed rice maki and soy sauce foam 26

Whisky: Irish coffee 68

White vinegar: Goat cheese flan, Dijon mustard caramel 62

White wine, dry: White sangria 74

Yeast, fresh: Fermented grape juice 128; Puffed rye bread 92

Yogurt: Cilantro-infused yogurt and carrot coulis 126

Index
of recipes